4th International Symposium on Atomic Technology 2009

(ISAT-4)

Journal of Physics: Conference Series Volume 232

Kobe, Japan
18-19 November 2009

ISBN: 978-1-61738-657-2
ISSN: 1742-6588

Printed from e-media with permission by:

Curran Associates, Inc.
57 Morehouse Lane
Red Hook, NY 12571

Some format issues inherent in the e-media version may also appear in this print version.

Printed by Curran Associates, Inc. (2010)

Institute of Physics
Dirac House, Temple Back
Bristol BS1 6BE UK

Phone: 44 1 17 929 7481
Fax: 44 1 17 920 0979

techtracking@iop.org

Additional copies of this publication are available from:

Curran Associates, Inc.
57 Morehouse Lane
Red Hook, NY 12571 USA
Phone: 845-758-0400
Fax: 845-758-2634
Email: curran@proceedings.com
Web: www.proceedings.com

TABLE OF CONTENTS

The International Symposium on Atomic Technology (ISAT) is held every year. The 4th Symposium (ISAT-4) was held on November 18-19, 2009 at the Seaside Hotel MAIKO VILLA KOBE, Kobe City, Japan presided by the "Atomic Technology Project".

The ISAT-4 symposium was intended to offer a forum for the discussion on the latest progress in the atomic technologies. The symposium was attended by 107 delegates. There were 10 invited and 6 oral presentations. The number of poster presentations was 69. From all the contributions, 22 papers selected through review process are contained in this volume.

The "Atomic Technology Project" was started in 2006 as a joint project of three institutions; (1) the Center for Atomic and Molecular Technologies, Graduate School of Engineering, Osaka University (CAMT), (2) the Tsukuba Research Center for Interdisciplinary Materials Science, Graduate School of Pure and Applied Sciences, University of Tsukuba (TIMS) and (3) the Polyscale Technology Research Center, Research Institute for Science and Technology, Tokyo University of Science (PTRC), each of which were independently pursuing nano-technologies and was developing atomic scale operation and diagnostics, functional materials, micro processing and device. The project is funded by the Ministry of Education, Culture, Sports, Science and Technology of Japan.

The goal of the project is to contribute to the development of atomic-scale science and technologies such as functional molecules, biomaterials, and quantum functions of atomic-scale structures.

Shigefumi Okada
Conference Chair
Center for Atomic and Molecular Technologies, Graduate School of Engineering, Osaka University, 2-1 Yamadaoka, Suita-city, Osaka 565-0871, Japan.

Application of liposome-encapsulated ceramic phoshpors for cancer cell imaging under near infrared excitation

Hirotada Akiyama[1,2], Kimikazu Tokuzen[2,3], Hiroko Otsuka[1], Kohei Soga[2,3], Fumio Tashiro[1,2]

[1] Department of Biological Science and Technology, Tokyo University of Science, [2] Polyscale Technology Research Center, Tokyo University of Science, [3] Department of Materials Science and Technology, Tokyo University of Science

hakiyama@rs.noda.tus.ac.jp

Abstract. Bioimaging with fluorescent probes is used as an invaluable tool in a biomedical field both *in vivo* and *in vitro*. However, organic dyes have some problems such as photo-breaching and cytotoxicity due to short wavelength with high quantum energy. Recently, a new approach using rare-earth-doped ceramic nanophosphors (RED-CNP) shows that fluorescence from RED-CNP in both visible (upconversion) and near infrared (NIR) wavelength region under NIR excitation is available for bioimaging. In order to efficiently introduce the RED-CNP into cancer cells, in this study we have developed a lipid nano-particles of liposome-encapsulated erbium (Er) ion-doped Y_2O_3 (lipo-Y_2O_3). Cationic lipo-Y_2O_3 could clearly visualize the intracellular region of human hepatocellular carcinoma Huh-7 cells by a fluorescence microscope measurements equipped with near-infrared excitation source scanning. The results imply that the lipo-Y_2O_3 would potentially be useful material for imaging of cancer cells. The embedded Y_2O_3 in the liposome having cancer-specific ligands and/or antibodies on its surface should have a great potential for cancer cell imaging in general in living subjects.

1. Introduction

Fluorescence bioimaging is an important technique to visualize phenomena and migration of substances in biological systems by using fluorescent agents. Many of the problems in fluorescence bioimaging, such as damages on biological substances, color fading, autofluorescence and light scattering are caused by the irradiation of ultraviolet (UV) or visible (VIS) light for excitation. The use of rare-earth-doped ceramic nanophosphors as a fluorescent agent can solve the above problems, because they are known to exhibit upconversion (visible) and efficient NIR fluorescence under NIR excitation [1]. Fluorescence bioimaging in the NIR range of 0.8 to 2.0 μm attracts a great deal of attention for the visualization of deeper lying images in living body without damage, because the use of the NIR light can reduce light scattering which is a major cause of the loss of both excitation and fluorescence lights in a living body *in vivo*. The emission intensity of the fluorescence of these nanophosphors does not decay. Therefore, the long term observation is possible. To apply them as biophotonic devices, the dispersion stability under physiological conditions, the control of specific interaction with a target substance and biocompatibility must be introduced to the probe. Liposome is bilayer vesicle structured by phospholipids that have similar composition of biological membrane, and show low toxicity and antigenicity. Also, both hydrophilic and hydrophobic agents can be encapsulated in it and a variety of targeting ligands can be introduced on its surface. In this study, cationic liposome encapsulating erbium (Er) ion-doped Y_2O_3-ceramic nanoparticle (lipo-Y_2O_3) were prepared from dipalmitoyl phosphatidylcholine (DPPC), stearylamine (SA) and cholesterol by using a complex emulsion method. NIR-to-visible emission was observed from the lipo-Y_2O_3 under 980-nm NIR excitation in the field of biological microscope. We evaluated the potential ability of lipo-Y_2O_3 on the cancer cell imaging by the detection of intracellular Y_2O_3.

2. Experimental Section

2.1. Preparation of PEGylated erbium ion-doped Y_2O_3 ceramic nanoparticles (PEG-Y_2O_3-CNP)

The erbium ion-doped Y_2O_3-CNP were prepared by calcinating precursors at 900°C for 3 h. The precursors, Y,Er(CO_3)OH, was precipitated using homogeneous precipitation. A mixture of urea and nitrates of yttrium and erbium was dissolved water as a starting solution. By heating solution up to 80°C, the urea decomposed into CO_3^{2-} and NH_4^+ as precipitants for the precursor [2]. The PEG-Y_2O_3-CNP was prepared as previous described [3].

2.2. Preparation of lipo-Y_2O_3

Liposomes of cationic were prepared by using complex emulsion method [4]. The DDPC (40 mol%), cholesterol (40 mol%) and SA (20 mol%) were diluted in chloroform solution. PEG-Y_2O_3-CNP were suspended to 5 mg/ml in HEPES buffer (20 mM HEPES pH7.4, 150 mM NaCl). Added PEG-Y_2O_3-CNP suspension in lipid solution and emulsified a mixture with sonicator to form a water in oil (w/o) type emulsion. The above emulsion was then added to HEPES buffer agitated with homogenizer to form a complex water in oil in water (w/o)/w-type emulsion. Residual chloroform was removed by stirring for 12 h. The emulsion were centrifuged at 20,000g for 20 min at 10°C and then washed twice and resuspended to 1 mg/ml in HEPES buffer.

2.3. Cell lines and culture

Human cancer cell lines, HeLa (cervical cancer cell line) and Huh-7 (liver cancer cell line) were obtained from the RIKEN Cell Bank. Human normal diploid cell line, Hs68 (foreskin cell line) were from the Human Science Research Resource Bank (HSRRB). Cells were cultured in Dulbecco's modified Eagle's medium (DMEM) supplemented with 10% fetal bovine serum (FBS) at 37°C under 5% CO_2 in a humidified atmosphere. To visualize the incorporated lipo-Y_2O_3, Huh-7 cells were seeded 1×10^4 cells/well in a Lab-Tek II Camber slide and cultivated for 24 h. Then the cells were treated with 0.05 mg/ml lipo-Y_2O_3 or PEG-Y_2O_3-CNP in serum free DMEM for 6 h at 37°C. Following the reagent treatments, the medium was exchanged to 10% FCS DMEM. After 18 h, the cells were washed in PBS, fixed in 4% paraformaldehyde. The cells were observed under a fluorescent microscope equipped with near-infrared excitation source (TCLDM9, Thorlabs, Newton, NJ).

2.4. Cytotoxicity assay

Normal and cancer cell lines were seeded 2×10^4 cells/well in 24-well plate and cultured for 24 h. Various doses of lipo-Y_2O_3 were added to cells and incubated for 48 h and cytotoxicity of lipo-Y_2O_3 was examined by trypan blue dye exclusion assay.

3. Results and Discussion

3.1. Observation of l lipo-Y_2O_3

Figure 1 shows scheme (a) and SEM images (b, c) of lipo-Y_2O_3 obtained by complex emulsion method. The resulting lipid vesicles were mostly spherical and unilamellar in appearance. The mean diameter of lipid vesicles determined by analysis of electron micrographs was about 500 ± 200 nm. PEG-Y_2O_3-CNPs of the size of about 200 nm from 5 to 10 were embedded in the liposome. Thus, we successfully prepared lipo-Y_2O_3

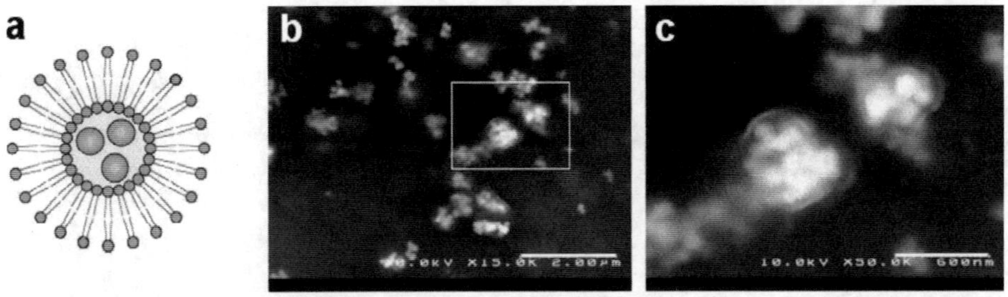

Figure 1. Scheme and SEM images of Lipo-Y_2O_3 .
(a) Scheme of Lipo-Y_2O_3 . (b, c) SEM images of Lipo -Y_2O_3 . Insert in (b) is shown in (c). Scale bars present 2 μm for (b) and 600 nm for (c).

3.2. Cytotoxicity of lipo-Y_2O_3

To investigate the cytotoxicity of lipo-Y_2O_3, human hepatoma Huh-7 cells, human cervical carcinoma HeLa cells and normal human foreskin cell line Hs68 cells were treated with various concentrations of lipo-Y_2O_3 for 24 h and their cell viabilities were assayed by trypan blue dye exclusion procedure. In both cancer and normal cells, severe damage caused by lipo-Y_2O_3 was not observed (Figure2), showing that this dose-range of lipo-Y_2O_3 using for bioimaging does not affect various cell viabilities.

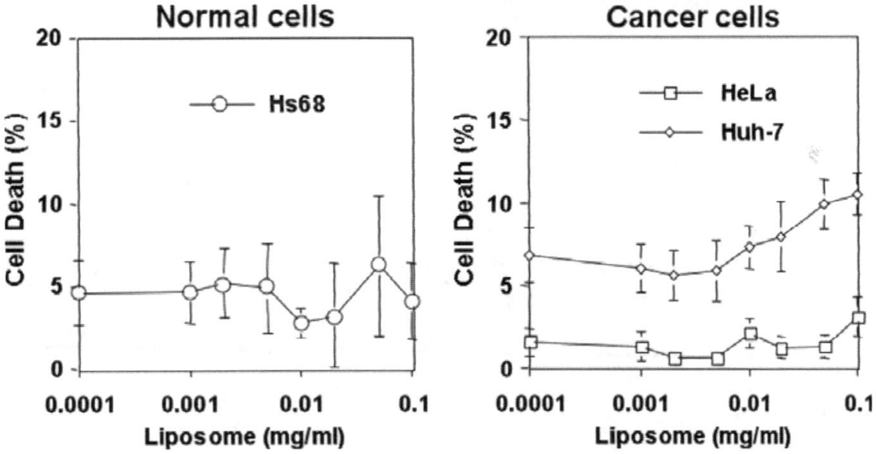

Figure 2. Cytotoxic effect of Lipo -Y_2O_3 on normal and cancer cells. Human normal and cancer cell lines were seeded 2 × 10^4 cells/well in 24 well plate and cultured in DMEM supplemented with 10% FCS for 24 h. The cells were incubated with various concentrations of Lipo -Y_2O_3 for 48 h and cytotoxicity of Lipo -Y_2O_3 was examined using trypan blue dye exclusion assay.

3.3. Detection of lipo-Y_2O_3 in intracellular region of cancer cell

To investigate the incorporation of PEG-Y_2O_3-CNP into cancer cells, Huh-7 cells were incubated with 0.05 mg/ml lipo-Y_2O_3 or PEG-Y_2O_3-CNP at 37°C for 6 h and observed under a fluorescence microscope equipped with NIR excitation source. As shown in Figure 3, NIR fluorescence was clearly detected as a dot in the intracellular region of lipo-Y_2O_3-treated cells, suggesting that lipo-Y_2O_3 was incorporated into the cells by endocytosis. On the other hand, based on the fluorescence pattern of PEG-Y_2O_3-CNPs, the CNPs were thought to be nonspecifically adsorbed on the cell surface. These results indicated that lipo-Y_2O_3, but not PEG-Y_2O_3-CNP, was efficiently incorporated into Huh-7 cells. In addition, efficient labeling of the cells was observed by the treatment with 0.01 - 0.1 mg/ml lipo-Y_2O_3. In this dose-range any cytotoxic effects were not detected (Figure 2).

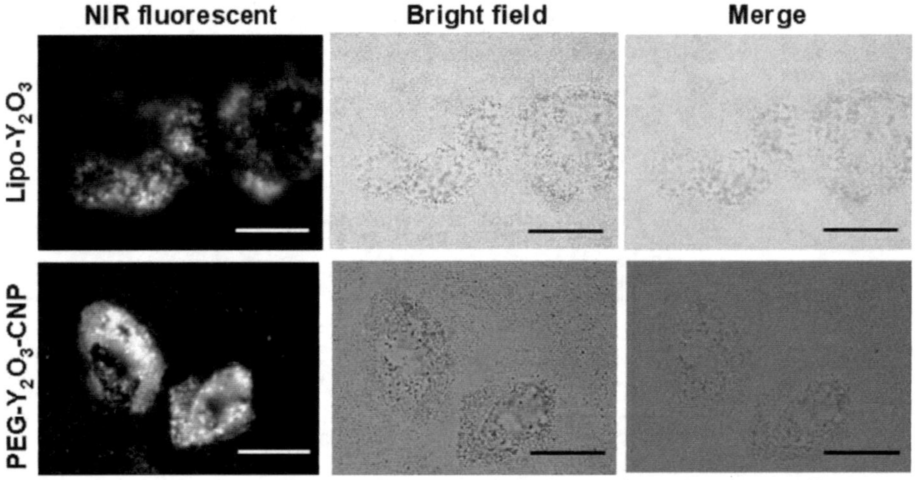

Figure 3. Detection of lipo-Y_2O_3 in intracellular region of cancer cells. Huh-7 cells were incubated with 0.05 mg/ml Lipo-Y_2O_3 or PEG-Y_2O_3-CNP at 37°C for 6 h. The cells were examined under a fluorescence microscope equipped with a 980 nm laser diode for the NIR excitation source. Scale bars present 30 μm.

4. Conclusion

The present data indicate that the lipo-Y_2O_3 might potentially be useful material for imaging of cancer cells. The NIR fluorescence imaging in living nude mice which are subcutaneously inoculated with Y_2O_3-CNP-labeled cancer cells leads to elucidation of the inversion and metastasis processes of malignant tumor cells. Moreover, functional modification of Lipo-Y_2O_3 by cancer-specific ligands and/or antibodies will confer a great potential for cancer cell imaging in general in living subjects. Furthermore, efficacious cancer therapy without any adverse effects will be enabled by enclosing an anticancer agent in liposome.

References

[1] Soga K. et al. 2003 *J. Appl. Phys.*, **93,** 2946-2951.
[2] Zako T. et al. 2008 *J. Mater. Sci.*, **43,** 5325–5330.
[3] Kamimura M. et al. 2008 *Langmuir,* **24,** 8864-8870.
[4] Fumiyoshi I. et al. 1988 *J. Dispersion. Sci. Tech.*, **9,** 1-15.

Inverse Spin Valve Effect in Multilayer Graphene Device

H Goto[1,4], **S Tanaka**[1,2,4], **H Tomori**[1,2,4], **Y Ootuka**[1,2], **K Tsukagoshi**[3,4], **and A Kanda**[1,4]

[1] Institute of Physics, University of Tsukuba, Tsukuba, Ibaraki 305-8571, Japan
[2] Tsukuba Research Center for Interdisciplinary Materials Science (TIMS), University of Tsukuba, Tsukuba 305-8571, Japan
[3] International Center for Materials Nanoarchitectonics (MANA), National Institute for Materials Science (NIMS), Namiki, Tsukuba, Ibaraki 305-0044, Japan
[4] Japan Science and Technology Agency (CREST-JST), Kawaguchi, Saitama 332-0012, Japan

E-mail: `hgoto@lt.px.tsukuba.ac.jp`

Abstract. We report the gate-voltage dependence of the spin transport in multilayer graphene (MLG) studied experimentally by the local measurement. The sample consists of a Ni/MLG/Ni junction, where the thickness of the MLG is 9 nm and the spacing of two Ni electrodes is 300 nm. At zero gate voltage, we observed the normal spin valve effect, in which the resistance for the antiparallel alignment of magnetization in ferromagnetic electrodes is larger than that for the parallel alignment. By applying a large gate voltage, on the other hand, the spin valve effect is reversed: the resistance for the antiparallel alignment becomes smaller than that for the parallel alignment. The result is qualitatively interpreted as a quantum interference effect, indicating that the mean free path and the spin relaxation length of the MLG are longer than the electrode spacing (300 nm).

1. Introduction

Spin transport in graphite-based materials such as carbon nanotubes [1–4], single-layer graphene (SLG) [5–7], and multilayer graphene (MLG) [8–11] has attracted considerable attention in recent years because coherent spin transport is expected over a long distance due to the weak spin-orbit and hyperfine interactions in these materials. Gate control of spin conduction in such materials is of high interest from the viewpoint of realizing multi-functional spintronic devices and clarifying the underlying physics. In particular, in view of practical applications, MLG should be superior to SLG because the influence of sample imperfections such as ripples on the spin and charge transports is expected to be smaller in MLG due to its stiffness [12], leading to longer mean free path l_e and spin relaxation length λ_N, although one needs to take into account the finite screening length of the gate electric field in MLG [13]. In this paper, we explore the gate controllability of the spin transport in MLG. We report the gate voltage dependent spin transport, indicating that the mean free path and the spin relaxation length are longer than the electrode spacing of 300 nm.

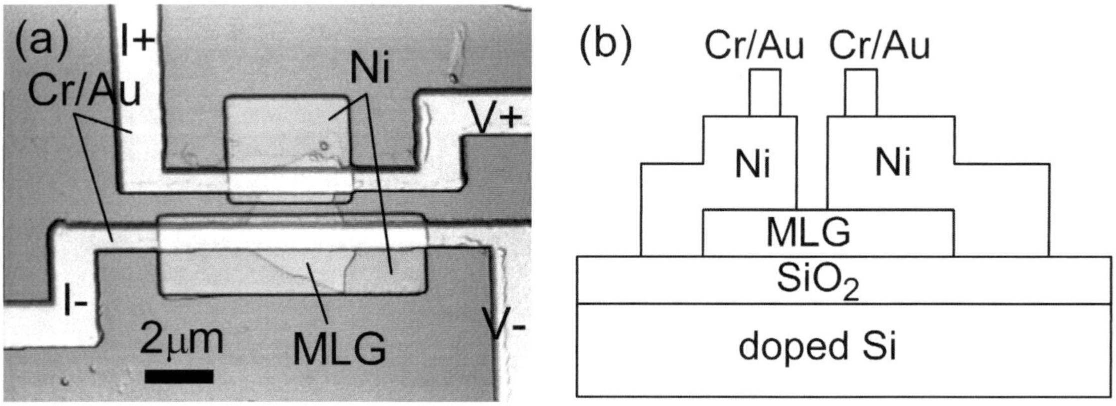

Figure 1. (a) Atomic force micrograph of the sample for the local measurement. Configuration and materials of electrodes are indicated. (b) Schematic side view of the sample.

2. Experimental

The samples were fabricated by the standard micromechanical cleavage technique of bulk graphite [14], followed by the electron beam lithography and metal deposition: a layer of MLG about 10 nm thick was exfoliated from a grain of kish graphite and placed on a SiO_2/highly doped Si substrate. Two ferromagnetic rectangles (Ni, 40 nm thick) with different aspect ratios were connected to the MLG. Then two normal-metal wires (Cr/Au, 5/100 nm thick) were attached to the Ni rectangles as the measurement leads. Figure 1 shows the atomic force micrograph and a schematic side view of the sample.

We examined more than ten samples with different MLG thicknesses and junction resistances. Here we report the results of a sample with 9nm thick MLG and a resistance of around 40 Ω (depending on the gate voltage), which is the smallest among all samples we have examined so far. In measurement, the output voltage V was measured using the four-terminal lock-in technique with an excitation current I of 10 μA and a frequency of 119 Hz. The in-plane magnetic field H was applied parallel to the long side of the ferromagnetic electrodes. Different aspect ratios of two Ni rectangles lead to different coercive forces, realizing the parallel and antiparallel alignment of the magnetization depending on the magnetic field. The back gate voltage was applied through the 300 nm thick SiO_2 layer. The sample was immersed in liquid ^{4}He and kept at 4.2 K. Thickness and sample geometry were determined with AFM and SEM after transport measurement.

3. Results

Figure 2 shows magnetoresistance (MR) curves measured at different gate voltages, $V_g = -80$, 0, and +80 V. The MR exhibits hysteresis corresponding to the parallel and anti-parallel alignment of the magnetization in ferromagnetic electrodes, as indicated in Fig. 2. Here, the variation of the MR is small and obscure in comparison with the conventional MR curves [6] presumably due to the resistance mismatch [15] at the transparent interface between Ni and MLG and due to the domain structures [16] in the wide Ni electrodes. Notice that at $V_g = 0$ the MR for antiparallel configuration of magnetization is larger than that for parallel configuration, corresponding to the normal spin valve effect commonly observed in ferromagnetic metal (FM)/normal metal/FM systems, while at $V_g = -80$ and 80 V the MR for antiparallel configuration is smaller than that for parallel configuration, i.e., the spin valve effect is reversed.

In order to analyze the effect quantitatively, we define spin induced magnetoresistance (SIMR)

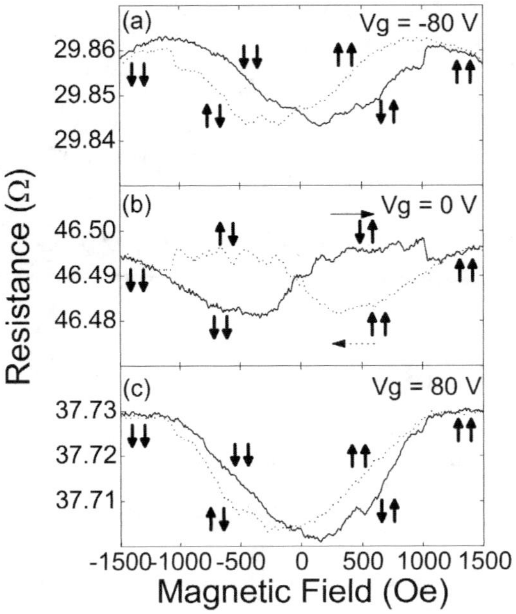

Figure 2. Magnetoresistance measured at V_g = -80, 0, 80 V from top to bottom. Solid (dashed) lines are MR in increasing (decreasing) magnetic field. Arrows indicate the magnetization configurations of ferromagnetic electrodes.

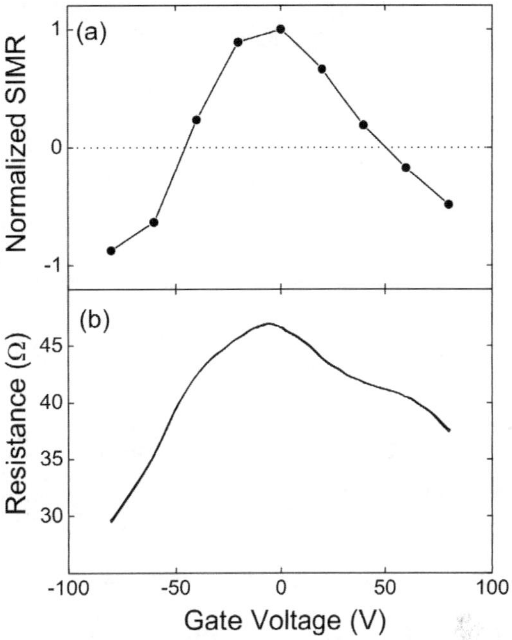

Figure 3. (a) Spin induced magnetoresistance (SIMR) and (b) resistance at zero field as a function of gate voltage. The SIMR is normalized so that the maximal value is equal to be 1.0.

as

$$\mathrm{SIMR} \equiv \int_{-H_c}^{0} [R_{down}(H) - R_{up}(H)]dH + \int_{0}^{H_c} [R_{up}(H) - R_{down}(H)]dH, \qquad (1)$$

where $H_c \approx 1200$ Oe is the larger coercive force in the two FMs, and $R_{up}(H)$ ($R_{down}(H)$) is magnetoresistance in increasing (decreasing) magnetic field. The calculated SIMR is shown in Fig. 3(a) as a function of the gate voltage. For comparison, the two-terminal resistance of the device measured at $H = 0$ is plotted in Fig. 3(b). Positive correlation between SIMR and resistance is clearly seen, indicating that the inverse spin valve effect is related to small device resistance.

4. Discussion

We now discuss the origin of the inverse spin valve effect. First, we remark that the effect is not originated from the extrinsic magnetoresistance such as anisotropic magnetoresistance or Hall effect, because these MRs depend only on magnetization alignment in ferromagnetic electrodes and are not affected by gate voltage or resistance. There are some candidates which reverse the spin valve effect; these are

1. the Rashba spin-orbit scattering [17],
2. the quantum interference [18] such as resonant states through Coulomb blockade [2], Fabri-Pérot interference [3] or wave vector matching [4], and
3. the ferromagnet-oxide interface [1, 10].

Using FM/graphitic material/FM junctions, several groups have reported the inverse spin valve

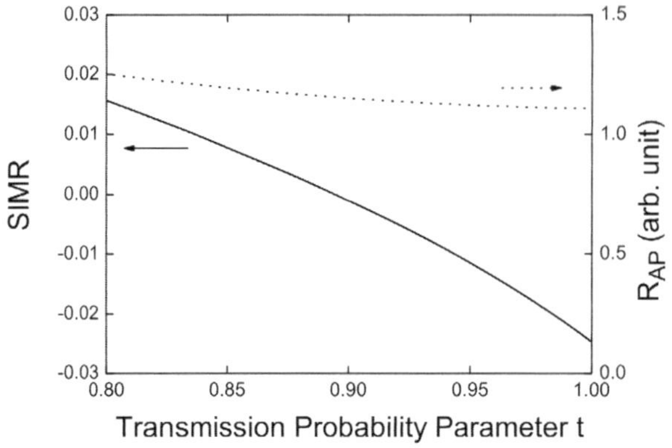

Figure 4. Calculated SIMR (solid line) and G_{AP} (dashed line) as a function of transmission probability parameter

effect due to not only gate voltage [2–4, 7, 10], but also temperature [1] and source-drain bias [9] and their results are attributed to the above mentioned origins.

Our results can be explained by the quantum interference, assuming that the transmission probabilities at the two interfaces increase with the gate voltage. In the Fabri-Pérot interference model [3], the transmission probability for the majority spin (+) and minority spin (-) of the left(s) and right (r) contacts is defined with

$$T_{sr} = T_L T_R | \sum_{n=0}^{N} [(1 - T_L^s)(1 - T_R^r)]^{n/2} e^{2in\delta} |^2, \tag{2}$$

where, following Refs. [2, 3], $T^s_{L(R)}$ is spin-dependent transmission probabilities at the left (right) contacts. Here, $T^s_{L(R)} = T_{L(R)}(1 + sP)$, $s = \pm 1$ and spin polarization P is assumed to be 0.23 [19]. The phase δ acquired by an electron at the round trip between two electrodes is fixed to be zero, and the repetition number N coherently reflected at the electrodes is set to be 2. We define the transmission probability parameter t with $T_L = 0.75t$ and $T_R = 0.25t$ and plot the SIMR $\equiv [(T_{+-} + T_{-+})^{-1} - (T_{++} + T_{--})^{-1}]/(T_{++} + T_{--})^{-1}$ and $R_{AP} \equiv (T_{+-} + T_{-+})^{-1}$ as a function of t in Fig. 4. The SIMR is positive at small t, but turns to negative with increasing t. R_{AP} is also a simply decreasing functions of t. Even though this calculation is oversimplified because the V_g dependence of δ and the number of conductance channels are neglected, the result is consistent with the correlation between the SIMR and the resistance shown in Fig. 3.

We emphasize that the quantum interference indicates the electrons and spins transport coherently in MLG, with l_e and λ_N substantially longer than the electrode spacing, 300 nm in the present sample. The mean free path of 300 nm implies the mobility of 12,000 $cm^2 \cdot V^{-1} \cdot s^{-1}$, which is easily attained in MLG at low temperatures. The long spin relaxation length is also consistent with our previous report [11] which shows λ_N much larger than 8.4 μm.

Finally, we estimate the *intrinsic* spin coherence length in MLG. Without magnetic impurities, the spin-orbit scattering is responsible for the spin relaxation. The intrinsic spin-orbit scattering time is given by $\tau_{so} = (\alpha Z)^{-4}\tau_e$ [20], where τ_e is the elastic scattering time, $\alpha = 1/137$ is the fine structure constant, and Z is the atomic number. The spin relaxation length in MLG is estimated to be $\lambda_N = (\alpha Z)^{-2} l_e \sim 500 l_e$. Substituting the typical value $l_e = 100$ nm, we obtain macroscopically long spin relaxation length $\lambda_N = 50$ μm. Thus, the light atomic mass and the large mobility make MLG a promising material for spintronics and related applications.

5. Conclusion

In conclusion, we succeeded in controlling the spin conduction in MLG by the electric field, which was confirmed by the local measurement. The inverse spin valve effect was observed by applying the large gate voltage. The results suggest the quantum coherence in MLG, which signifies that the mean free path and the spin relaxation length in MLG are longer than electrode spacing, 300 nm. The gate tuning of MR opens the new possibility of graphite-based spintronic devices and MLG is a promising material for its realization.

Acknowledgments

This work is partly supported by the Grants-in-Aid for Scientific Research (17069004) from the Ministry of Education, Culture, Sports, Science and Technology (MEXT) of Japan, and the Inter-university Cooperative Research Program of the Institute of Materials Research, Tohoku University.

References

[1] Chakraborty S, Walsh K M, Alphenaar B W, Liu L and Tsukagoshi K 2003 *Appl. Phys. Lett.* **83** 1008

[2] Sahoo S, Kontos T, Furer J, Hoffmann C, Gräber M, Cottet A and Schönenberger C 2005 *Nature Phys.* **1** 99

[3] Man H T, Wever I J W and Morpurgo A F 2006 *Phys. Rev. B* **73** 241401(R)

[4] Nagabhirava B, Bansal T, Sumanasekera G U, Alphenaar B W and Liu L 2006 *Appl. Phys. Lett.* **88** 023503

[5] Hill E W, Geim A K, Novoselov K, Schedin F and Blake P 2006 *IEEE Trans. Magn.* **42** 2694

[6] Tombros N, Jozsa C, Popinciuc M, Jonkman H T and van Wees B J 2007 *Nature (London)* **448** 571

[7] Cho S, Chen Y F and Fuhrer M S 2007 *Appl. Phys. Lett* **91** 123105

[8] Ohishi M, Shiraishi M, Nouchi R, Nozaki T, Shinjo T and Suzuki Y 2007 *Jpn. J. Appl. Phys.* **46** L605

[9] Nishioka M and Goldman A M 2007 *Appl. Phys. Lett* **90** 252505

[10] Wang W H, Pi K, Li Y, Chiang Y F, P Wei J S and Kawakami R K 2008 *Phys. Rev. B* **77** 020402(R)

[11] Goto H, Kanda A, Sato T, Tanaka S, Ootuka Y, Odaka S, Miyazaki H, Tsukagoshi K and Aoyagi Y 2008 *Appl. Phys. Lett.* **92** 212110

[12] Morozov S V, Novoselov K S, Schedin F, Jiang D, Firsov A A and Geim A K 2005 *Phys. Rev. B* **72** 201401

[13] Miyazaki H, Odaka S, Sato T, Tanaka S, Goto H, Kanda A, Tsukagoshi K, Ootuka Y and Aoyagi Y 2008 *Applied Physics Express* **1** 034007

[14] Novoselov K S, Geim A K, Morozov S V, Jiang D, Zhang Y, Dubonos S V, Grigorieva I V and Firsov A A 2004 *Science* **306** 666

[15] Schmidt G, Ferrand D, Molenkamp L W, Filip A T and van Wees B J 2000 *Phys. Rev. B* **62** R4790

[16] Wernsdorfer W, Hasselbach K, Mailly D, Barbara B, Benoit A, Thomas L and Swan G 1995 *J. Magn. Magn. Mater.* **145** 33

[17] Datta S and Das B 1990 *Appl. Phys. Lett* **56** 665

[18] Schäpers T, Nitta J, Heersche H B and Takayanagi H 2001 *Phys. Rev. B* **64** 125314

[19] Meservey R and Tedrow P M 1994 *Phys. Rep.* **238** 173

[20] Abrikosov A A and Gorkov L P 1962 *Sov. Phys. JETP* **15** 752

Giant vortex and multivortex states under local supercurrent injection in a mesoscopic superconducting square

S Hatsumi[1,2], **Y Kuroda**[1,2], **Y Ootuka**[1,2], **A Kanda**[1]

[1]Institute of Physics, University of Tsukuba, Tsukuba 305-8571, Japan
[2]Tsukuba Research Center for Interdisciplinary Materials Science (TIMS), University of Tsukuba, Tsukuba 305-8571, Japan

E-mail: hatsumi@lt.px.tsukuba.ac.jp

Abstract. We experimentally study giant vortex states (GVSs) and multivortex states (MVSs) in a mesoscopic superconducting square under local supercurrent injection. By using the multiple-small-tunnel-junction measurement, we find that the magnetic field for the transition between a GVS and an MVS with a fixed vorticity depends on the magnitude of the injected supercurrent. This result indicates the possibility of manipulating quantum vortex states with supercurrent injection.

1. Introduction

Mesoscopic superconductors have sizes comparable to the superconducting coherence length ξ and/or the magnetic penetration length λ. In such small superconductors, confinement of supercurrent dramatically modifies superconducting properties. In particular, vortex configuration strongly reflects the sample geometry, leading to formation of novel vortex states such as multi vortex states (MVSs) and giant vortex states (GVSs). In MVSs, repulsive interaction between a vortex and the shielding supercurrent flowing along the sample boundary forces the vortex arrangement to fit the sample geometry. On the other hand, when the shielding current is large enough and/or the sample size is small enough, the resultant strong confinement effect stabilizes the GVS, in which multiple singly-quantized vortices merge into a multiply-quantized giant vortex. These two novel vortex states were theoretically predicted for a long time [1, 2, 3, 4, 5, 6], but have been experimentally confirmed just recently.[7, 8, 9, 10]

It should be noted that in most of the experimental studies reported so far, transitions between different vortex states have been induced by magnetic field sweep, which is not so convenient for the manipulation of the vortex states because fast sweep of magnetic field is accompanied by a large eddy current. In this study, we try to induce vortex state transitions by applying local supercurrent into the superconductor. Some of the main results are already published elsewhere.[11, 12] Here we report transitions between giant vortex and multivortex states with a fixed vorticity L (the change of the phase of the order parameter around the sample divided by 2π) under local supercurrent injection.

In order to distinguish between GVSs and MVSs, we have developed the multiple-small-tunnel-junction (MSTJ) method.[7, 13] In a sample for the MSTJ measurement, multiple small

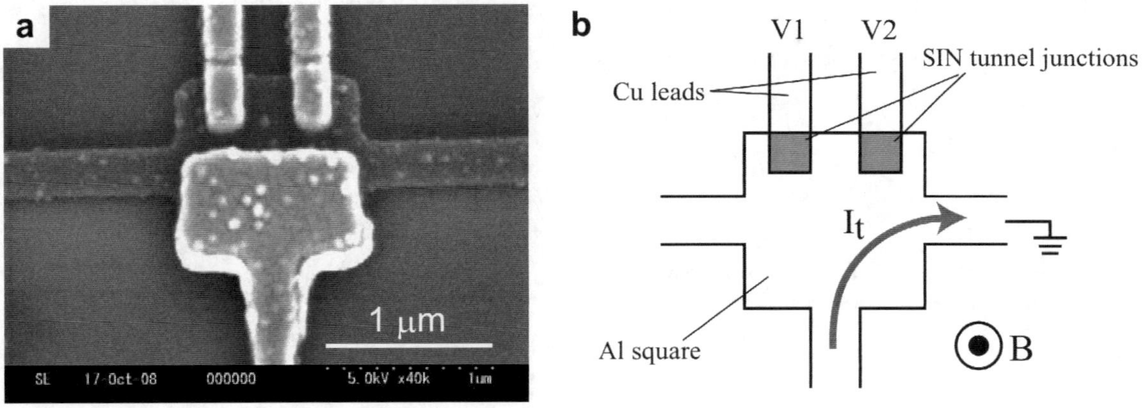

Figure 1. (a) An SEM image and (b) a schematic view of the sample. The sample is Al superconducting square with two SIN junctions and three leads. The side and thickness of Al square are 1.1 μm and 40 nm.

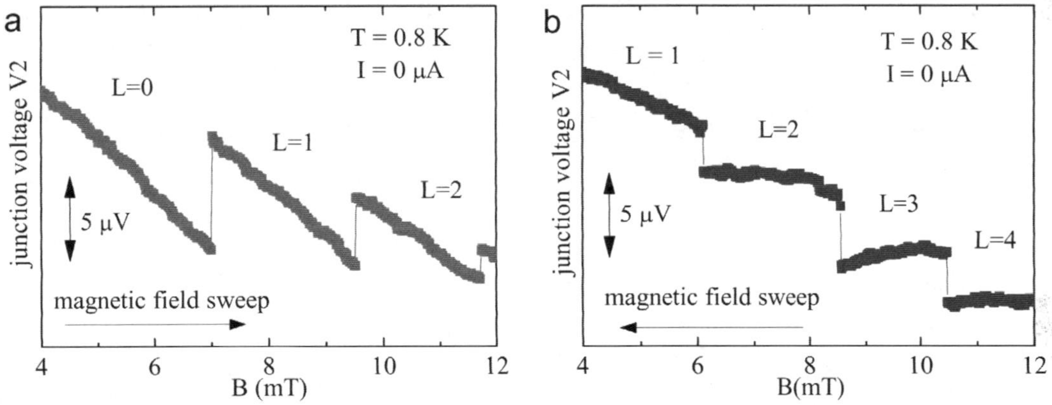

Figure 2. Magnetic field dependence of junction voltage $V2$ in (a) increasing and (b) decreasing magnetic field at 0.8 K.

superconductor-insulator-normal metal (SIN) junctions are attached to symmetrical positions of a mesoscopic superconductor. It is well known that when a small current is applied to the SIN junction, the measured voltage reflects the superconducting energy gap underneath the junction, which changes sensitively with the local supercurrent density. So, by comparing the junction voltages, one can obtain (partial) information about the supercurrent distribution, which is closely related to the vortex configuration. Particularly, when the voltage of each junction placed at symmetrical positions take the same value, the vortex state is potentially the GVS, while when they are different to each other, the vortex state is definitely an MVS.[7]

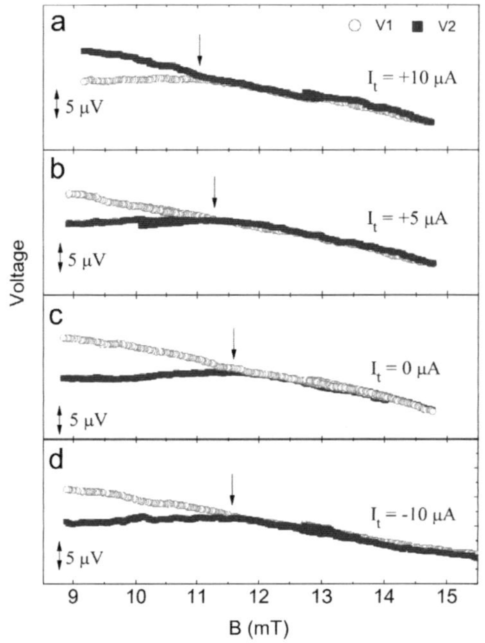

Figure 3. Magnetic field dependence of junction voltages $V1$ and $V2$ in the $L - 3$ state for injection supercurrent $I_t =$ (a) 10μA, (b) 5μA, (c) 0μA, (d) -10μA at 0.7 K.

2. Measurements and Results

The sample is an Al square with side of 1.1 μm and thickness of 40 nm, which was fabricated by using e-beam lithography followed by the angle deposition of Al and Cu. Figure 1 shows a scanning electron microscope (SEM) image and a schematic view of the sample. Two Cu leads are connected to symmetrical positions in the top side of the sample through Al-AlO_x-Cu SIN tunnel junctions. In addition, three Al leads are attached to the center of the sides directly. In the measurement, the bottom lead is used for the current injection and the right lead is connected to the ground. The left lead is left unconnected. The sample was cooled in a dilution refrigerator. A fixed current of 1 nA is injected to each SIN junction, and the junction voltages, $V1$ and $V2$, were measured simultaneously as a function of the perpendicular magnetic field. We estimated the coherence length $\xi(0)$ to be $150 - 190$ nm and the superconducting critical temperature T_c was 1.35 K.

Figure 2 shows the typical magnetic field dependence of the junction voltage. There are two origins for the voltage change: one is related to the smearing of the gap structure in the density of states by magnetic field and the other is the decrease of the superconducting energy gap by supercurrent flowing underneath the junction.[14] Especially, a large voltage jump corresponds to a drastic change of supercurrent, presumably due to vortex penetration/expulsion. We are able to determine the vorticity L, as shown in Fig. 2, by taking into account the fact that vortices penetrate into (or, is expelled from) the sample one by one.

Figure 3 shows the junction voltages $V1$ and $V2$ for the whole $L = 3$ state at 0.7 K under several injection currents, $I_t = 10\mu$A, 5 μA, 0 μA, and -5 μA. The directions of the positive injection current and the magnetic field are indicated in Fig. 1(b). We compensated the difference of the offset voltages of the voltage amplifiers in the measurement system by assuming that the voltages $V1$ and $V2$ take the same values at the highest magnetic field of the $L = 3$ state, i.e., the vortex state is the GVS close to the vortex penetration. This assumption for $I_t = 0$ (Fig. 3(c)) is justified by the numerical simulation based on the Ginzburg-Landau theory.[12] In Fig. 3(c), we notice that $V1$ and $V2$ takes almost the same value above $B = 11.6$ mT (indicated by an arrow), while their difference becomes wider with decreasing magnetic field below 11.6 mT. Here, we can distinguish between MVS and GVS by comparing between junction voltages $V1$ and $V2$ in the case of $L = 3$, $I_t = 0$ [9]; in GVSs, a giant vortex is situated at the center of the sample, leading to the quadruple symmetry in the distribution of supercurrent. As a result, voltages of junctions attached to symmetric positions of the sample, $V1$ and $V2$, are equal to each other. On the other hand, in the $L = 3$ MVSs, vortex configuration does not have such quadruple symmetry, so $V1$ and $V2$ take different values. For this reason, we can guess that vortex state is the GVS at high magnetic fields ($B > 11.6$ mT) and an MVS at low magnetic field ($B < 11.6$ mT). Thus an MVS-GVS transition occurs at 11.6 mT. Here we note that both the $V1$ curve and the $V2$ curve have a kink at 11.6 mT, indicating the change of the vortex structure.

In the case of $I_t \neq 0$, the symmetry of supercurrent distribution is broken by the injected supercurrent even in GVSs, so strictly speaking, we can't use the MSTJ method for the distinction between MVSs and GVSs. However, the behavior seen in Figs. 3(a), 3(b) and 3(d) is similar to that of Fig. 3(c), i.e., $V1$ almost equals $V2$ above a magnetic field at which clear kinks are seen both in $V1$ and $V2$, suggesting the GVS in this magnetic field range. The magnetic field for the MVS-GVS transition is not necessarily the same as that of Fig. 3(c): 11.0 mT, 11.3 mT, and 11.6 mT in Figs. 3(a), 3(b) and 3(d), respectively, as indicated by arrows. These results demonstrate that the MSTJ measurement is still valid under supercurrent injection as shown in Fig. 3, although the core of a GVS is expected to be shifted from the sample center due to the Lorentz force exerted by the injected supercurrent and the symmetry of the supercurrent distribution is supposed to be broken. Notice that when the magnetic field is fixed at a value between 11.0 mT and 11.6 mT, an MVS-GVS transition is expected under the sweep of the injected supercurrent I_t, which is consistent with the observation by the temperature dependence of the vortex expulsion fields.[8, 9, 12] Also note that in Fig. 3(a), $V2$ is larger than $V1$ in the MVS, while in other cases (Figs. 3(b), 3(c) and 3(d)), $V2$ is smaller than $V1$ in the MVS. This shows that the vortex configuration at $I_t = 10\mu$A is different from that in $I_t < 5\mu$A. Such a kind of MVS-MVS transitions at a fixed vorticity is discussed in more detail elsewhere.[12]

3. Conclusion

We studied vortex states in a mesoscopic superconducting square under local supercurrent injection by using the MSTJ method. We find that the MVS-GVS transition occurs at the $L = 3$ state, and the magnetic field at which the MVS-GVS transition occurs depends on the injected supercurrent. The result supports the existence of the MVS-GVS transition as a function of the locally injected supercurrent. This result indicates the possibility of manipulating quantum vortex states with supercurrent injection.

Acknowledgments

This work was supported by the Kurata Memorial Hitachi Science and Technology Foundation. AK greatly acknowledges M. V. Milošević, F. M. Peeters, and M. Hayashi for discussions.

References

[1] Moshchalkov V V, Qiu X G and Bruyndoncx V 1997 *Phys. Rev. B* **55** 11793

[2] Schweigert V A, Peeters F M and Deo P S 1998 *Phys. Rev. Lett.* **81** 2783.

[3] Palacois J J 1998 *Phys. Rev. B* **58** (1998) 5948(R)
[4] Palacois J J 2000 *Phys. Rev. Lett.* **84** 1796
[5] Baelus B J and Peeters F M 2002 *Phys. Rev. B* **65** 104515
[6] Baelus B J, Cabral L R E and Peeters F M 2004 *Phys. Rev. B* **69** 064506
[7] Kanda A, Baelus B J, Peeters F M, Kadowaki K and Ootuka Y 2004 *Phys. Rev. Lett.* **93** 257002
[8] Baelus B J, Kanda A, Peeters F M, Ootuka Y and Kadowaki K 2005 *Phys. Rev. B* **71** 140502(R)
[9] Baelus B J, Kanda A, Shimizu N, Tadano K, Ootuka Y, Kadowaki K and Peeters F M 2006 *Phys. Rev. B* **73** 024514
[10] Grigorieva I V, Escoffier W, Misko V R, Baelus B J, Peeters F M, Vinnikov L Y and Dubonos S V 2007 *Phys. Rev. Lett.* **99** 147003
[11] Hatsumi S, Ootuka Y and Kanda A 2009 *Physica C* **469** 1080
[12] Miloševič M V, Kanda A, Hatsumi S, Peeters F M and Ootuka Y 2009 *Phys. Rev. Lett.* **103** 217003
[13] Kanda A and Ootuka Y 2004 *Physica C* **404** 205
[14] Tinkham M 1996 *Introduction to Superconductivity 2nd ed.* (New York: McGraw-Hill) chapter 10

Formation of Transparent Aluminum Hydroxide Film with Mesoscopic Surface Roughness by Hydrothermal Treatment of Incompletely-nitrided Sputtered Aluminum Film

Toshiyuki Hori, Zhiyong Qiu and Takashi Ishiguro[1]
Department of Materials Science and Technology, Tokyo University of Science, 2641, Yamazaki, Noda, Chiba 278-8510, Japan

E-mail: ishiguro@rs.noda.tus.ac.jp

Abstract. Incompletely-nitrided Al films (Al-N film) are deposited on the glass substrate by rf sputtering with a metallic Al target and using Ar and N_2 gas mixture. With increasing film thickness up to 300nm, the surface roughness increases. And the roughness is easily controlled. The size and the number density of surface protuberance are suitable to control diffusive optical properties in the visible and near infrared regions. The films become transparent with retained roughness by boiling in ultra pure water at 368K under atmospheric pressure. The films have been transformed from composite of Al and AlN to aluminum hydroxide (Boehmite). Total transmittance of the boiled specimens exceeded that of the glass substrate itself. These facts suggest that hydrothermally-treated Al-N films with the mesoscopic surface roughness have high potential to reduce the optical loss by reflection.

1. Introduction

Full-fledged use of environmentally friendly energy, especially use of solar energy has been starting to prevent greenhouse warming. In the first place, solar energy is one of the energy resources with thin energy density. Therefore, we have to improve not only efficiency of solar cell itself but also reduction of optical loss due to reflection from the cell, which occurs essentially at surface and/or interface with abrupt change in refractive index.

In order to realize a structure of a kind of optical-matching, we use MgF[1] and CaF[2] etc., because their refractive index is around 1.4, which is one of the median values of 1.6 of glass and 1 of air. Or more specifically, surface with mesoscopic roughness plays a role of the gradient-refractive index. We have reported that the surface structure of Al and/or Al-N films is one of proper candidates [3-5]. Recently, we succeeded in transforming of Al and Al-N films from opaque to transparent one, which is useful for anti-reflective coating [6, 7]. This is quite simple method, i.e., just boiling the films in the ultra pure water, by which Al and Al-N films transform into aluminium hydroxide film.

In this study, we examine the controllability of the surface roughness for the Al-N films. Then, evaluation of transparent films by boiling has been performed systematically.

---

2. Experimental

Al-N films were deposited on 0.70mm-thick Corning 7059 glass substrate by using rf sputtering method. An Al target (99.999%, 100mm diameter) was sputtered in the mixture gas of Ar (99.9995 %) and N_2 (99.9995 %) with a concentration of 7.3 vol%N_2. The working gas pressure was 1.33 Pa. The input rf power was 200W. The substrate temperature (T_{sub}) was 373K. Film thickness (d_f) was changed from 100nm to 300nm.

As deposited Al-N films were boiled in stirred ultrapure water at 368K under atmospheric pressure. Ultrapure water with a resistivity of higher than 18.0 MΩ· cm was used. Boiling time was chosen from 2 min to 20 min depending on the film thickness.

Surface morphology was observed using an atomic force microscope (AFM: Nanopics, NPX100, Seiko Instrument Inc.). The optical properties were measured using an ultraviolet-visual-near infrared spectrometer (UV-3100PC, Shimazdu Co.) in the wavelength range of 0.24 to 2.60 μm. Specular reflectance (R_0) was evaluated by measuring the reflectance at an incidence angle of 12°. Diffusive reflectance (R_{dif}) was measured using an integrating sphere with a diameter of 60mm. The incident light energy (=1) is the sum of the total transmittance (T_{tot}), total reflectance (R_{tot}) and absorptance (A). T_{tot} is the sum of forward transmittance (T_0) and diffusive transmittance (T_{dif}). R_{tot} is the sum of R_0 and R_{dif}. Then, the rule of energy conservation is represented by $1= T_{tot} + R_{tot} + A = T_0 + T_{dif} + R_0 + R_{dif} + A$. In order to distinguish the incidence direction to a specimen having the film / substrate structure, the superscript of fil for the incidence on film surface or the superscript of sub for the incidence on substrate will be used, if necessary.

3. Results and Discussion

Figure 1 shows the difference in appearance of as-deposited Al-N films with different thickness and their change by boiling in water. With increasing film thickness, metallic surface (Fig. 1(a)) becomes blackened surfaces (Figs. 1(b) and (c)). After boiling, they change into transparent films as shown in Figs. 1(d), (e) and (f).

Optical properties of as-deposited opaque Al-N films are shown in Fig. 2. R_0 decreases with d_f. On the other hand, R_{dif} increases with d_f at longer wavelength region. And the peak position of R_{dif} seems to be shifted toward longer wavelength with d_f. Then, R_{tot} decreases with d_f. These facts mean that thicker Al-N film shows excellent light absorption properties in the visible and near infrared region [4].

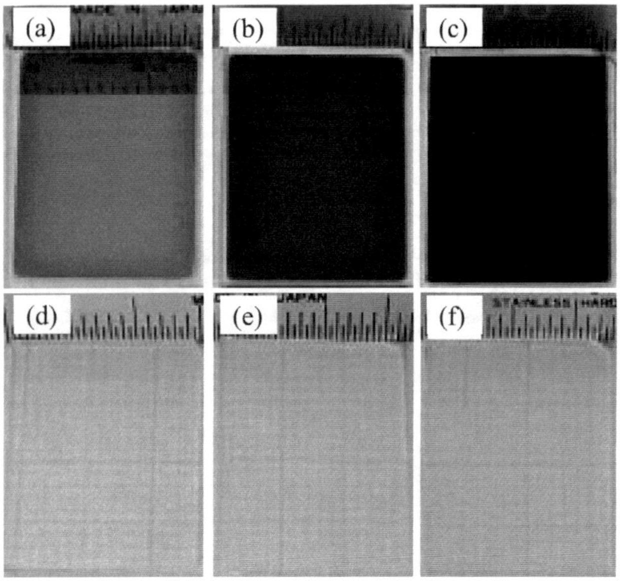

Figure 1. Appearance of as-deposited Al-N films with different film thickness on Corning #7059 glass substrate. (T_{sub} = 373K) (a) d_f=100nm. (b) 200nm. (c) 300nm. Each film becomes transparent (d), (e), and (f) after boiling at 368K for 2 min, 9 min, and 20 min, respectively.

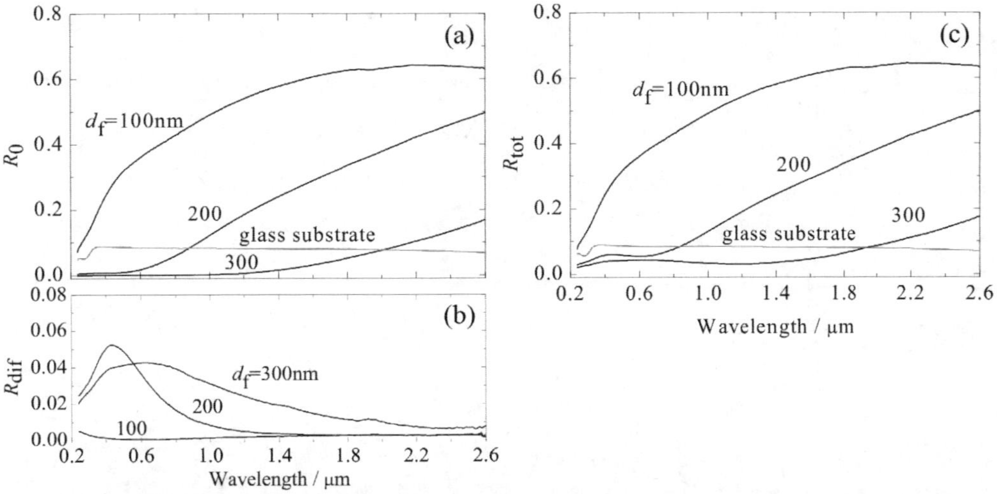

Figure 2. Film thickness dependence of specular reflectance (R_0) (a), diffusive reflectance (R_{dif}) (b) and total reflectance (R_{tot}) (c) for as-deposited Al-N films.

Optical properties of boiled specimens are shown in Fig. 3. As seen in Fig. 3(a), T_0 decreases with d_f at around visible region. T_{tot} of the boiled specimens exceeded that of the glass substrate itself (Fig. 3(b)). Then, T_{dif} is estimated by assuming zero absorption (A=0). As seen in Fig. 3(a) again, T_{dif} increases to compensate the depression of T_0. In Fig. 3(c) and (d), R_0, R_{dif} and R_{tot} corresponding to Figs. 1(d), (e), and (f) are also shown. With increasing d_f, R_0 decreases and R_{dif} increases. Then, R_{tot} falls below values of glass substrate. These optical properties, i.e., higher T_{tot} and lower R_{tot}, are quite useful property for the AR coating.

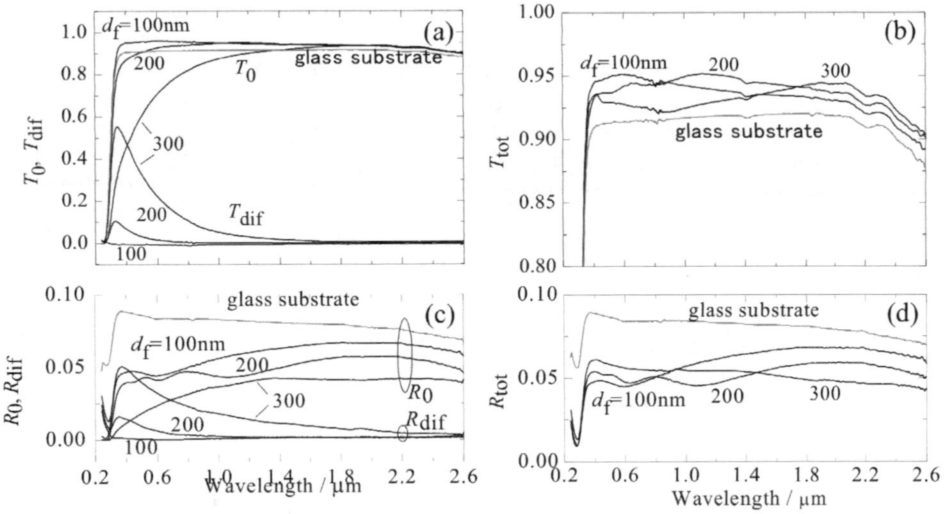

Figure 3. Film thickness dependence of (a) specular transmittance (T_0) and diffusive transmittance (T_{dif}), (b) total transmittance (T_{tot}), (c) specular reflectance (R_0) and diffusive reflectance (R_{dif}), and (d) total reflectance (R_{tot}). For comparison, optical properties of glass substrate are also shown.

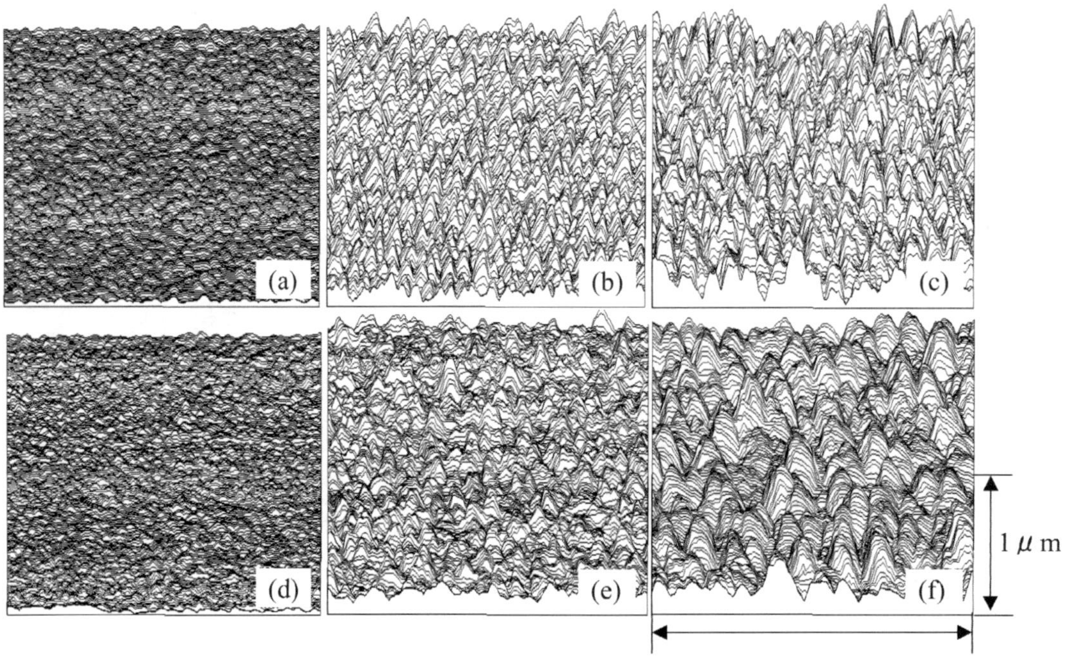

Figure 4. Bird's eye view AFM images of as-deposited (a) 100nm-thick, (b) 200nm-thick, and (c) 300nm-thick specimens and those of after boiling ((d), (e), and (f)). Here, (d), (e), and (f) corresponds to (a), (b), and (c).

AFM observation of the change in the surface morphology by boiling is shown in Fig. 4. Surface roughness increases with d_f (Figs. 4(a)-(c)). After boiling, surface roughness changes from Figs. 4(a), (b), and (c) into Figs. 4(d), (e), and (f). A tendency of increasing surface roughness with d_f can be seen. In addition, the width of each protuberance seems to grow.

In order to characterize the surface morphology, the root-mean-square (RMS) roughness and the areal number density of surface protuberance (N_{pro}) are estimated as shown in Fig. 5 from AFM images of Figs. 4. In the case of RMS roughness, difference between before and after boiling is not so large. It increases linearly with the film thickness for $d_f > 100$nm. In other words, surface roughness of as-deposited film is rapidly enhanced above d_f=100nm. On the otherhand, N_{pro} is seen to decrease by a factor of about 2, which means protuberance's growth in lateral direction takes place. Then, it could be mention that the degree and/or size of surface roughness can be controlled by initial Al-N film thickness.

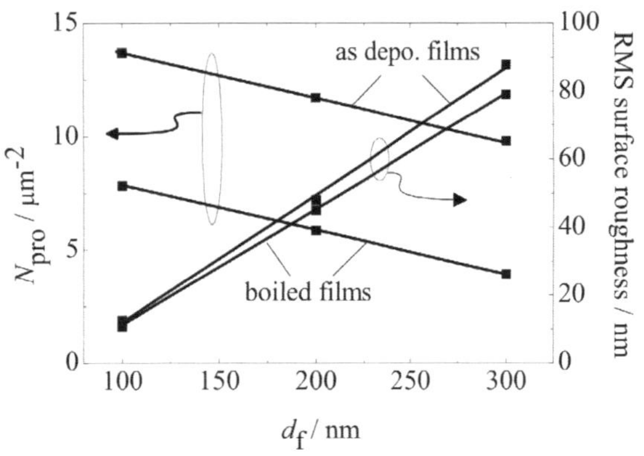

Figure 5. Film thickness dependence of the root-mean-square (RMS) surface roughness and the areal number density of surface protuberance (N_{pro}). Comparison between as-deposited film and boiled film is also shown.

Finally, complex refractive indices of *n-ik* is calculated by using optical data observed for boiled 30nm-thick specimen, which is specially-deposited at 293K to get a flat surface. Obtained values are almost same as the values for boiled Al film, which is described in our former paper [7]. Therefore, it is mentioned that *n* of boiled Al-N film or Boehmite film also shows intermediate values between glass and air. This is thought to be another reason for low reflectance.

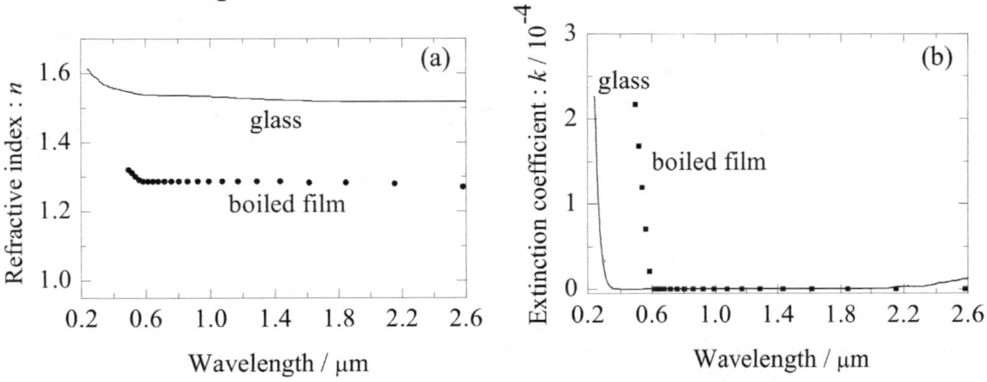

Figure 6. Calculated complex refractive indices of *n-ik* for boiled Al-N film and Corning #7059 glass.

4. Conclusion

We confirmed that the Al-N films with mesoscopic surface roughness were reformed to transparent films by hydrothermal treatment. Their surface roughness of the boiled Al-N film could be controlled by d_f of as deposited Al-N film. After hydrothermal treatment, with increasing surface roughness, T_0 decreases and T_d increases. However, T_{tot} for all of the specimens increases beyond T_{tot} of the glass substrate itself in all of wavelength ranges.

Complex refractive indices (*n-ik*) of the boiled Al-N film were calculated. Its refractive index (*n*) is around 1.3, which is one of the median values of 1.6 of glass and 1 of air.

These results indicate that the boiled Al-N film with mesoscopic surface roughness is useful for the AR coating such as solar cell and/or other optical devices.

Acknowledgments

One of authors (T. I.) was partially supported by a Grant-in-Aid for Scientific Research from the Ministry of Education, Culture, Sports, Science and Technology of Japan. This work was also supported by "Academic Frontier" Project for Private Universities: Matching Fund Subsidy from MEXT (Ministry of Education, Culture, Sports, Science and Technology), 2006-2010.

References

[1] Bendow B, Gianino P D, Tsay Y F and Mitra S S 1974 *Appl. Opt.*, **13**(10) 2382

[2] Lindan P J D and Gillan M J J 1993 *Phys. Condensed Matter.* **5** 1019

[3] Ishiguro T and Miyamura K 2005 *Mater. Trans.* **46**(12) 3037

[4] Ishiguro T, Nishimura M and Yamazaki T 2002 *Jpn. J. Appl. Phys.* **41,** 292

[5] Ishiguro T and Iijima H 2006 *Jpn. J. Appl. Phys.* **45** 7784

[6] Ishiguro T, Hattori D and Hamasaki K 2007 *Proceedings of ISES Solar World Congress 2007*

[7] Ishiguro T, Hori T and Qiu Z 2009 *J. Appl. Phys.* **106**, 23524

Transport properties of thermoelectric $SrTiO_3$ synthesized by polymerized complex method and spark plasma sintering

Mikio Ito[1] and Naoto Ohira[2]
[1] Center for Atomic and Molecular Technologies, Graduate School of Engineering, Osaka University
[2]Graduate Student, Osaka University
2-1 Yamadaoka, Suita, Osaka 565-0871, Japan

E-mail: ito@mat.eng.osaka-u.ac.jp

Abstract. The establishment of rapid synthesis process at a low temperature of thermoelectric non-doped and Y-doped $SrTiO_3$ was tried by using the polymerized complex process and the spark plasma sintering method with addition of TiB_2 powder. During the sintering, any reaction between $SrTiO_3$ phase and TiB_2 powder did not occur, resulting in the formation of $Sr_{1-x}Y_xTiO_3/TiB_2$ composites. The TiB_2 addition was quite effective not only for reduction in the electrical resistivity, but also for rapid densification during sintering. The power factor of the $Sr_{0.9}Y_{0.1}TiO_3/TiB_2$ composite was similar to that of the $Sr_{0.9}Y_{0.1}TiO_3$ polycrystal synthesized by the conventional solid state reaction method.

1. Introduction

Several electronic conductive oxides, such as $(Zn_{1-x}Al_x)O$, $BaSrPbO_3$, $Na_xCo_2O_4$ etc., have been recognized as potential candidates for a new thermoelectric material [1-6]. These thermoelectric oxides can be used at high temperatures without deterioration of their performance due to oxidation, and their production costs are comparatively low. The p-type thermoelectric oxides, such as $Na_xCo_2O_4$ and $Ca_3Co_4O_9$ etc., are well known to show the high thermoelectric performance [1,7-9]. On the other hand, the thermoelectric properties of the n-type oxides, such as $(Zn_{1-x}Al_x)O$ and $SrTiO_3$ etc., are still low as compared to those of the p-type oxides, and further improvement of the thermoelectric performance of the n-type oxides is required.

In the case of the n-type perovskite $SrTiO_3$ phase, the $SrTiO_3$ polycrystalline sintered body is generally synthesized by the conventional solid state reaction (SSR) method [10-12]. In order to improve the thermoelectric performance, rare-earth metal, such as La or Y is partially substituted for Sr site, or Nb is substituted for Ti site [12-14]. In the case of this SSR method, mother materials, $SrCO_3$, TiO_2, La_2O_3 or Y_2O_3 etc., are required to be subjected to calcination at a high temperature around 1673 K ~ 1873 K, pulverization and mixing repeatedly prior to the final sintering. Besides that, the sintering process is also carried out at a high temperature around 1873 K in Ar atmosphere by using hot pressing, etc. In this study, in order to establish a synthesis process of Y-doped $SrTiO_3$ at a lower temperature and for a shorter period of operating time, the polymerized complex (PC) process and the spark plasma sintering (SPS) process were adopted. The PC process is one of the chemical solution processes, and can provide a fine powder precursor and mix constituent metal elements at atomic level [15-17]. Therefore, the PC process can provide Y-doped $SrTiO_3$ powders at a lower temperature and for a shorter period of operating time than the SSR process [18]. The SPS process is well known as a method of rapid consolidation by applying pulsed high current to a powder sample. Additionally, TiB_2 powder, which is a good electrical conductive ceramic, is added to the $SrTiO_3$ powders on the purpose of reduction on electrical resistivity of the sintered samples. Thus, in this study, the synthesis of $SrTiO_3$ and Y-doped $SrTiO_3$ polycrystals with good thermoelectric performance was tried by the PC and SPS processes and TiB_2 addition. The effects of TiB_2 addition on their thermoelectric properties and sintering behavior during SPS were investigated.

2. Experimental procedure

$SrTiO_3$ thermoelectric oxides were synthesized from a powder precursor prepared by the PC process. $Ti[OCH(CH_3)_2]_4$ (95 % purity) was first dissolved in ethylene glycol (99.5 % purity) and then heated and stirred on a hot stirrer set at 473 K. Subsequently, citric acid ($C_5H_8O_7$) (98 % purity), $Sr(NO_3)_2$ (98 % purity), and $(CH_3COO)_3Y \cdot 6H_2O$ (99.9 % purity) were added to this solution. $Sr(NO_3)_2$, $Ti[OCH(CH_3)_2]_4$ and $(CH_3COO)_3Y \cdot 6H_2O$ were mixed in the mole ratio corresponding to nominal composition of $SrTiO_3$ or $Sr_{0.9}Y_{0.1}TiO_3$. Citric acid and ethylene glycol were mixed in the proportions of 4 moles and 150 moles for each mole of metal cation, respectively. The mixture was then stirred at 473~723 K. During this heating process, the formation of the polymer between ethylene glycol and metal citrate complexes was promoted. As the colloidal solution was condensed, it became highly viscous, and this viscous polymeric product was decomposed to a dark mass precursor at 723 K in air by a mantle heater. This mass precursor was ground and calcined at 1073 K for 36 ks in order to obtain the $SrTiO_3$ phase. In the case of the $Sr_{1-x}Y_xTiO_3/TiB_2$ composites, the calcined powder precursors were mixed with 10 mass% TiB_2 powder (1~2 μm in particle size) by using a mortar and pestle. 2 g of the powder mixture was put into carbon dies (inner diameter: 10 mm, outer diameter: 30 mm, hight: 30 mm). Then, these powders and powder mixtures were sintered via the SPS apparatus (SPS Syntex Inc., SPS-511S) under a pressure of 40 MPa at 1473 K for 5 min in vaccum. The temperature was determined by using a pyrometer and measuring the surface temperature of carbon dies during sintering. The phases and the microstructures of these samples were determined by XRD (X-ray diffraction analysis), SEM (scanning electron microscopy). The XRD analysis was performed on the pressed plane of the sintered compacts using Cu K_α radiation. The Seebeck coefficient, S, and the electrical resistivity, ρ, were simultaneously measured from room temperature to about 1170 K by the ordinary four probe dc method in a flowing Ar gas atmosphere using computer-controlled equipments, and then the power factor, P, of these samples were calculated using the measured S and ρ values in the equation $P=S^2/\rho$.

3. Results and discussions

Figure 1 shows the XRD patterns of the $SrTiO_3$ and $Sr_{0.9}Y_{0.1}TiO_3$ powders after PC process and the subsequent calcination. All the peaks of the non-doped sample were identified as those from the perovskite $SrTiO_3$ phase. The $Sr_{0.9}Y_{0.1}TiO_3$ powder was also found to be composed of the perovskite $SrTiO_3$ phase without any other peaks. Besides that, each peak of this powder was slightly broader than those of the non-doped sample. This is considered to be due to the lattice strain induced by Y doping. Although there was no significant difference between lattice constants of these powders, from the results of the XRD analysis, it is considered that the doped Y atoms were substituted for Sr site in the $SrTiO_3$ perovskite structure. These powders were mixed with 10 mass% TiB_2 powder and then subjected to the SPS process at 1473 K for 5 min. Figure 2 shows the XRD patterns of the non-doped and Y-doped $SrTiO_3$ polycrystals with TiB_2 addition. In the case of the non-doped and Y-doped $SrTiO_3$ without TiB_2 addition, both the samples showed the peaks from perovskite $SrTiO_3$ phase only. When TiB_2 powder was added, as shown in figure 2, these sintered bodies were composed of $SrTiO_3$ and TiB_2 phases without precipitation of another phase. Besides that, the relative densities of all the polycrystals obtained in this study were about 99 %. Thus, it was found that the densely sintered bodies of the $SrTiO_3/TiB_2$ and $Sr_{0.9}Y_{0.1}TiO_3/TiB_2$ composites could be obtained without reaction of added TiB_2 to $SrTiO_3$ phase, by the SPS process with TiB_2 powder addition.

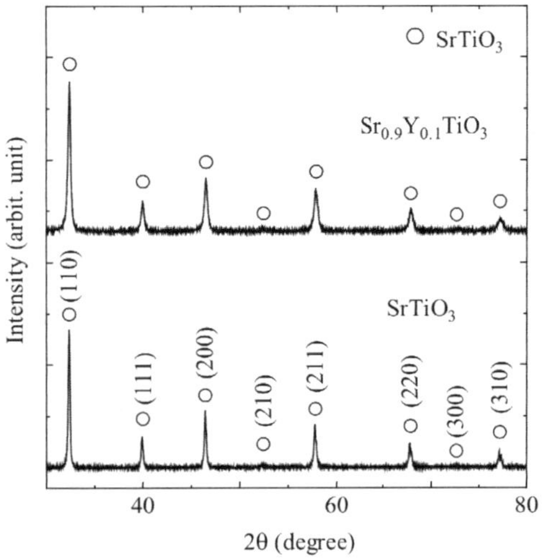

Figure 1. XRD patterns of $SrTiO_3$ and $Sr_{0.9}Y_{0.1}TiO_3$ powders synthesized by the PC process.

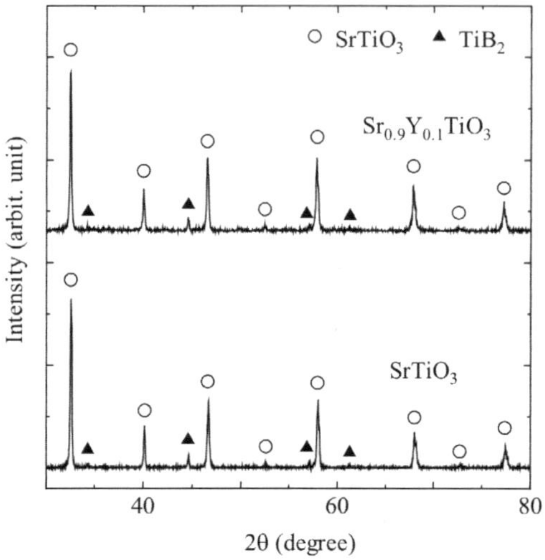

Figure 2. XRD patterns of the $SrTiO_3$ and $Sr_{0.9}Y_{0.1}TiO_3$ polycrystals with 10 mass% TiB_2 after the SPS process.

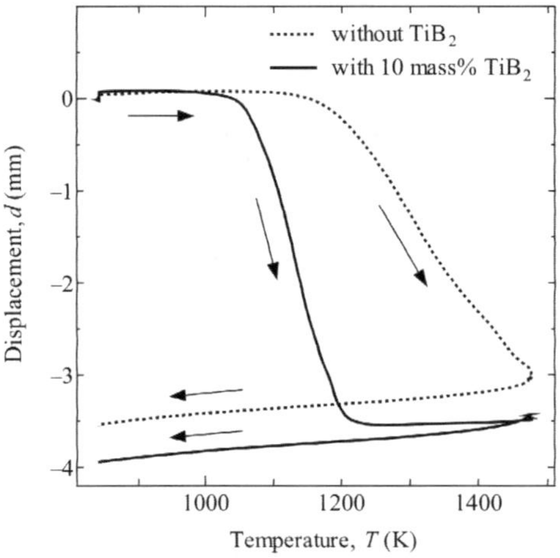

Figure 3. Temperature dependence of the displacement of an upper punch during the SPS process for the $Sr_{0.9}Y_{0.1}TiO_3$ powders with and without 10 mass% TiB_2.

Figure 3 shows the temperature dependence of the displacement of an upper punch measured by the SPS apparatus during sintering of $Sr_{0.9}Y_{0.1}TiO_3$ powders with and without TiB_2 addition. The downward displacement of the upper punch corresponds to densification of the sample during SPS process. In the case of the sample without TiB_2, the densification began at around 1150 K and last up to 1473 K during the temperature elevation, and then shrank gradually with decreasing temperature. On the other hand, when TiB_2 powder was added, the sample started shrinking at around 1030 K, which is by 120 K lower than that of the sample without TiB_2. Additionally the densification rate between 1080 K and 1200K was significantly large as compared to that of the non-doped sample

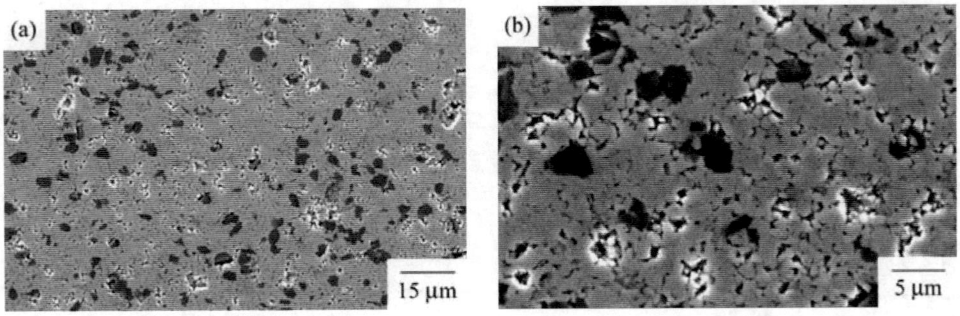

Figure 4. SEM photographs of (a) the cross section of the $Sr_{0.9}Y_{0.1}TiO_3$ polycrystals with 10 mass% TiB_2 after the SPS process and (b) the enlarged one from the photograph (a).

between 1200 K and 1470 K, and the sample with TiB_2 addition almost completed its densification at around 1250 K. In the case of the non-doped $SrTiO_3$ sample, it was also confirmed that densification was markedly accelerated by TiB_2 addition and finished at around 1250 K, which is almost the same behaviour as that of the Y-doped sample as shown in figure 3. Although the mechanism causing the low temperature densification and the large shrinkage rate has not been clarified yet, it was found that the TiB_2 addition was quite effective on accelerating densification of $SrTiO_3$ ceramics during the SPS process.

Figure 4 shows the SEM photographs of (a) the cross section of Y-doped $SrTiO_3$ polycrystal with TiB_2 addition and (b) enlarged one from the photograph (a). As shown in figure 4 (a), the sample was composed of the gray $SrTiO_3$ matrix and the black TiB_2 particles. This result of SEM observation is well consistent with that of the XRD analysis shown in figure 2. The morphology of the $SrTiO_3$ matrix was almost the same as that of the $Sr_{0.9}Y_{0.1}TiO_3$ without TiB_2 addition. The TiB_2 phase mostly maintained its original particle size around 1~2 μm after the sintering, and dispersed uniformly in the $SrTiO_3$ matrix. The interfaces between the $SrTiO_3$ matrix and the TiB_2 dispersoids were clear, and another phase formation, such as reaction of these two phases, were not detectable, as can be seen in figure 4 (b). Generally, reduction in densification temperature and/or acceleration of shrinkage rate is expected to be possible to occur when a reaction between constituent phases or a liquid phase formation, etc., contributes to sintering process. Therefore, such a rapid densification is considered to be difficult to occur just in the case of the simple mixing of a matrix phase and an additive phase. In this study, as described in figures 2 and 4, there were not any detective reaction products from the constituent phases and any sign of liquid phase formation. Thus, the $SrTiO_3$ matrix phase and the TiB_2 dispersoids seemed to be form a simple composite, and no microstructural factor causing the rapid densification due to TiB_2 addition shown in figure 3 could be observed. In the SPS process, consolidation process generally proceeds more rapidly by applying pulsed high current to a sample powder, as compared to the conventional sintering processes, such as pressureless sintering, hot pressing, etc. When a sample with a large electrical resistivity is sintered via SPS process, the pulsed high current mainly flows along a carbon die and the sample was sintered by joule heating of the carbon die (when the sample temperature becomes high, the pulsed current also begins to flow through the compact.). On the other hand, in the case of the sample with a small electrical resistivity, it is considered that the pulsed high current flows across the powder sample directly from a lower temperature and densification of the compact is accelerated by joule heating on the interparticle area in the compact. The non-doped $SrTiO_3$ sample without TiB_2 addition synthesized in this study was electrically insulate-like. Even when Y was doped, the electrical resistivity of the sintered sample was quite large, so the Seebeck coefficient and electrical resistivity of these samples without TiB_2 addition could not be measured by using our measurement apparatus. On the other hand, the electrical resistivity of these sintered samples was significantly reduced by addition of 10 mass% TiB_2, and the

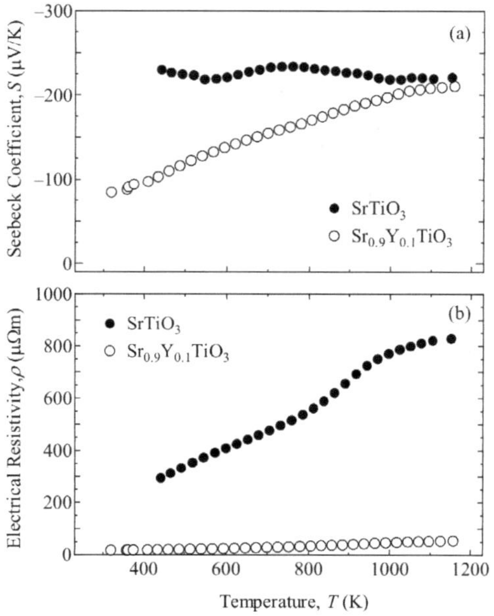

Figure 5. Temperature dependence of (a) the Seebeck coefficient, *S*, and (b) the electrical resistivity, *ρ*, of the $SrTiO_3$ and $Sr_{0.9}Y_{0.1}TiO_3$ polycrystals with 10 mass% TiB_2 after the SPS process.

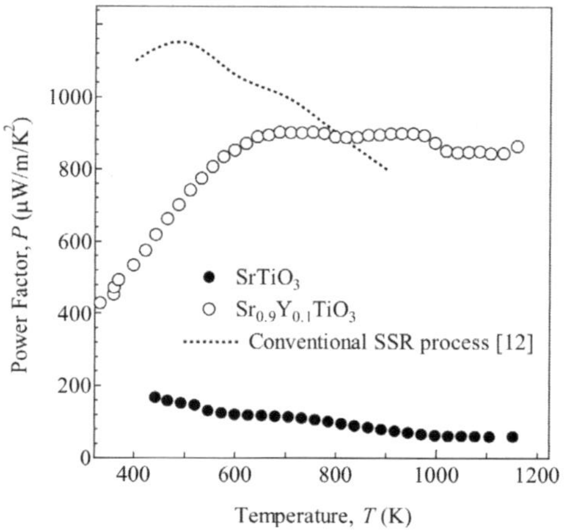

Figure 6. Temperature dependence of the power factor, *P*, of the $SrTiO_3$ and $Sr_{0.9}Y_{0.1}TiO_3$ polycrystals with 10 mass% TiB_2 after the SPS process, and with the *P* value of the $Sr_{0.9}Y_{0.1}TiO_3$ synthesized by the conventional SSR method [12].

electrical transport properties of these samples with TiB_2 addition could be evaluated, as described below. Thus, the TiB_2 addition provided the big difference of electrical resistivity of the $SrTiO_3$ compacts, which is suggested to result in the rapid densification observed in figure 3.

Figure 5 shows the temperature dependence of (a) the Seebeck coefficient, *S*, and (b) the electrical resistivity, *ρ*, of the non-doped and Y-doped $SrTiO_3$ polycrystals with 10 mass% TiB_2. The Seebeck coefficient of the non-doped $SrTiO_3$ was around 220~230 μV/K. On the other hand, the electrical resistivity of this sample increased with increasing temperature and reached 830 μΩm at 1150 K. In the case of the Y-doped sample, an increase in carrier concentration due to Y doping reduced the *S* values and markedly decreased the *ρ* values over the entire measured temperature range, as compared to those of the non-doped sample. From the results shown in figures 1 and 5, it is supported that the doped Y atoms were substituted for Sr site and acted as an n-type dopant.

Figure 6 shows the temperature dependence of the power factor, P ($=S^2/\rho$), of the non-doped and Y-doped $SrTiO_3$ polycrystals with 10 mass% TiB_2. The power factor of the $Sr_{0.9}Y_{0.1}TiO_3$ sample synthesized by the conventional SSR process [12] was also illustrated for comparison in the figure. The non-doped sample showed the power factor less than 200 $\mu W/m/K^2$ and the *P* values decreased monotonously with increasing temperature, because of the significant increase in the electrical resistivity. On the other hand, Y doping greatly improved the power factor over the entire measured temperature range, which resulted from the significant reduction in the *ρ* values. The *P* values of the Y-doped sample remained around 900 $\mu W/m/K^2$ on the wide range between 680 K and 970 K. The lower temperature range below 800 K, the SSR sample showed *P* values larger than the $Sr_{0.9}Y_{0.1}TiO_3$ polycrystal with 10 mass% TiB_2 obtained in this study. On the other hand, above 800 K, the power factor of the sample prepared in this study exceeded that of the SSR sample. Thus, it was found that the Y-doped $SrTiO_3$ thermoelectric materials could be synthesized quite efficiently (at a lower temperature and for a shorter period of operating time), as compared to the conventional SSR process.

It is expected that further improvement of the thermoelectric properties may be possible, if the synthesis condition, such as a sintering temperature, a sintering time, an amount of added TiB_2, etc., are optimized in the future.

4. Conclusions

The $Sr_{1-x}Y_xTiO_3/TiB_2$ (x=0 and 0.1) composites were synthesized by mixing of $SrTiO_3$ phase powder produced by the PC process and TiB_2 powder, and the subsequent SPS method. The addition of TiB_2 powder, which is a good electrical conductor, caused the significant decrease in the starting temperature of densification and acceleration of shrinkage rate during SPS process. The electrical resistivity of the sintered polycrystals was also greatly reduced by the TiB_2 addition. The $Sr_{0.9}Y_{0.1}TiO_3/TiB_2$ composite sintered body showed the power factor similar to that of the $Sr_{0.9}Y_{0.1}TiO_3$ sample synthesized by the SSR method. From the results obtained in this study, the powder preparation by the PC process and the consolidation by the SPS method with addition of TiB_2 powder were found to be quite effective for the rapid synthesis process at a lower temperature of Y-doped $SrTiO_3$ thermoelectric materials, as compared to the conventional SSR process.

References

[1] Ohtaki M, Tsubota T, Eguchi K and Arai H, 1996 *J. Appl. Phys.* **79** 1816-18

[2] Yasukawa M and Murayama N, 1997 *J. Mater. Sci. Lett.* **16** 1731-34

[3] Terasaki I, Sasago Y and Uchinokura K, 1997 *Phys. Rev. B* **56** R12685-87

[4] Ohtaki M and Maeda E, 2000 *J. Jpn. Soc. Powder Powder Metall.* **47** 1159-64

[5] Nagira T, Ito M, Katsuyama S and Hara S, 2003 *Mater. Trans.* **44** 1866-71

[6] Ito M and Nagira T, 2005 *Mater. Trans.* **46** 1456-61

[7] Ito M and Furumoto D, 2007 *Mater. Trans.* **48** 3160-63

[8] Funahashi R, Matsubara I, Ikuta H, Takeuchi T, Mizutani U and Sodeoka S, 2000 *Jpn. J. Appl. Phys.* **39** L1127-29

[9] Katsuyama S, Takiguchi Y and Ito M, 2007 *Mater. Trans.* **48** 2073-78

[10] Moos R, Gnudi A and Hardtl K H, 1995 *J. Appl. Phys.* **78** 5042-47

[11] Hashimoto S, Kindermann L, Poulsen F W and Mogensen M, 2005 *J. Alloy. Compd.* **397** 245-49

[12] Obara H, Yamamoto A, Lee C H, Kobayashi K, Matsumoto A and Funahashi R, 2004 *Jpn. J. Appl. Phys.* **43** L540-42

[13] Okuda T, Nakanishi K, Miyasaka S and Tokura Y, 2001 *Phys. Rev. B* **63** 113104.

[14] Ohta S, Nomura T, Ohta H, Hirano M, Hosono H and Koumoto K, 2005 *Appl. Phys. Lett.* **87** 092108.

[15] Kakihana M, Yoshimura M, Mazaki H, Yasuoka H and Borjesson L, 1992 *J. Appl. Phys.* **71** 3904-10

[16] Durán P, Capel F, Tartaj J and Moure C, 2002 *J. Euro. Ceramic. Soc.* **22** 67-77

[17] Ito M, Nagira T, Furumoto D, Katsuyama S and Nagai H, 2003 *Scripta Mater.* **48** 403-08

[18] Ito M and Ohira N, 2008 *Mater. Trans.* **49** 1844-47

Reduction in densification temperature by TiB_2 addition during sintering of oxides

Mikio Ito[1], Naoki Hasegawa[2] and Takayoshi Nakano[3]
[1] Center for Atomic and Molecular Technologies, Graduate School of Engineering, Osaka University
[2]Graduate Student, Osaka University
[3]Division of Materials and Manufacturing Science, Graduate School of Engineering, Osaka University
2-1 Yamadaoka, Suita, Osaka 565-0871, Japan

E-mail: ito@mat.eng.osaka-u.ac.jp

Abstract. Several oxide powders ($SrTiO_3$, $BaTiO_3$ and Y_2O_3) were mixed with TiB_2 powder and then sintered by the SPS process. The effects of TiB_2 addition on the densification behaviours of these oxides during SPS were investigated and its mechanisms were discussed. The TiB_2 addition was found to be quite effective for reducing the densification temperature of these oxides during sintering. In the case of the $SrTiO_3$, after sintering, the sample was composed of $SrTiO_3$ and TiB_2 only, indicating that reactive sintering and/or liquid phase sintering did not occur. Even when sintered by the conventional hot pressing, it was found that TiB_2 addition was also quite effective for reduction in densification temperature as well as when sintered by SPS, suggesting that the effect of TiB_2 addition is available for the conventional heating process. These experimental results show the possibility of TiB_2 as a sintering agent for low temperature densification of oxides.

1. Introduction

The perovskite type $SrTiO_3$ has a high melting point, 3253 K and is practically used as a ceramic condenser. Recently, $SrTiO_3$ has also been a potential candidate for an n-type thermoelectric material [1-5]. However, consolidation to a polycrystal from an oxide powder with a high melting point generally requires a sintering process at high temperature and for a long time. On the other hand, it is well known that the SPS (spark plasma sintering) process is quite effective for consolidating powder oxides, powder metals etc., at a lower temperature and for a shorter period of operating time as compared to the conventional sintering process such as pressureless sintering, hot pressing and so forth. In the synthesis process of the n-type thermoelectric $SrTiO_3$ system, we discovered that the sintering by the SPS process with TiB_2 powder addition is effective for reduction in electrical resistivity, resulting in enhancement of thermoelectric performance, and for decrease in densification temperature during the SPS process [6]. It should be noted that the addition of around 10 mass% TiB_2 can reduce the densification temperature of a $SrTiO_3$ compact by up to 200 K. Thus, TiB_2 addition is considered to be quite important for preparation process at lower temperature and for a short period of operating time in a practical manufacturing process. However, the mechanism of the low temperature densification process provided by TiB_2 addition has not been clarified yet. Besides that, this rapid sintering process at a low temperature may also be available for consolidation process of other oxide powders. Based on these considerations, in this study, we investigated the possibility of application of TiB_2 addition to the sintering of several other oxides, and tried to clarify the origin of the reduction in densification temperature.

2. Experimental procedure

A commercial $SrTiO_3$ powder (Wako Chemicals, Nano-sized) was mixed with 10 mass% TiB_2 powder (Waco Chemicals, 2-4μm). 1 g of the powder mixture was put into carbon dies (inner size: 15 mm×7mm, outer diameter: 30 mm, hight: 30 mm). Then, the powder mixture was sintered by using SPS apparatus at 1573 K for 5 min under 55 MPa in vacuum. The temperature was determined by using a pyrometer and measuring the surface temperature of carbon dies during sintering. The sintering behaviour was evaluated by measuring shrinkage of the powder compact during SPS process. The sintering behaviours of other oxides, $BaTiO_3$ powder (Wako Chemicals, Nano-sized, 98% purity), Y_2O_3 powder (Nano Teck, mean particle diameter: 20.6-88.5 nm, 99.5% purity) and Al_2O_3 powder (Nano Teck, mean particle diameter: 31 nm, 99.9% purity) etc., were also observed in the same conditions. In order to investigate the mechanism of reduction in the densification behaviour caused by the SPS process with TiB_2 addition, the powder mixture of $SrTiO_3$ and TiB_2 was also sintered by the conventional hot pressing (HP) and its shrinkage curve was compared to that of the sample sintered by SPS. Additionally, $SrTiO_3$ powder was also mixed with 10 mass% Cu powder, and then its sintering behaviour was observed in the same condition and compared to that of the mixture sample of $SrTiO_3$ and TiB_2.

The phases and the microstructures of these samples were determined by XRD (X-ray diffraction analysis), SEM (scanning electron microscopy). The XRD analysis was performed on the pressed plane of the sintered compacts using Cu K_α radiation.

3. Experimental Results

Figure 1 shows the temperature dependence of displacement of the upper punch for $SrTiO_3$ samples with and without 10 mass% TiB_2 during SPS process. The displacement of the upper punch corresponds to shrinkage of a powder compact during sintering, and the effect of thermal expansion of the upper and lower punches was subtracted. The main shrinkage of the $SrTiO_3$ sample without TiB_2 addition started at about 1110 K, and lasted up to 1350 K. On the other hand, in the case of the sample with 10 mass % TiB_2, the densification occurred around 1100 K and finished at 1200 K. Although the final densification of these samples was almost the same, it was found that the densification

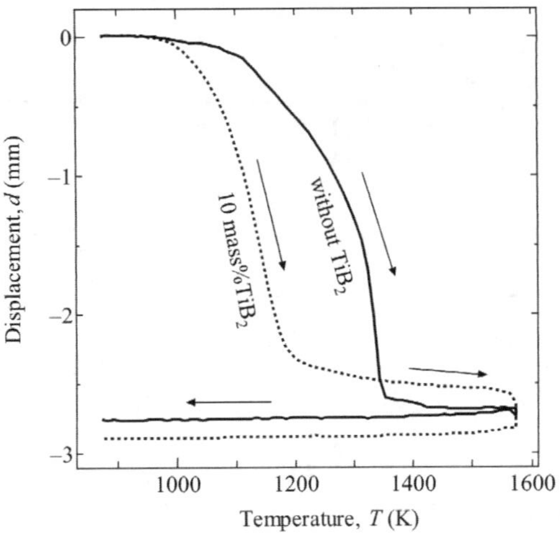

Figure 1. Densification behaviours of $SrTiO_3$ with and without 10 mass% TiB_2 during SPS.

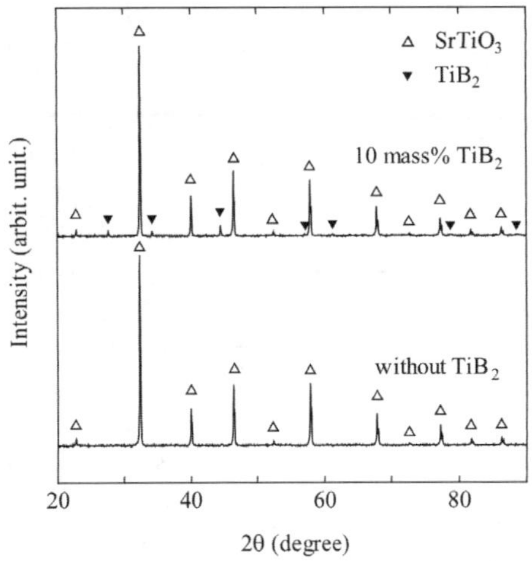

Figure 2. XRD patterns of $SrTiO_3$ with and without 10 mass% TiB_2 after SPS.

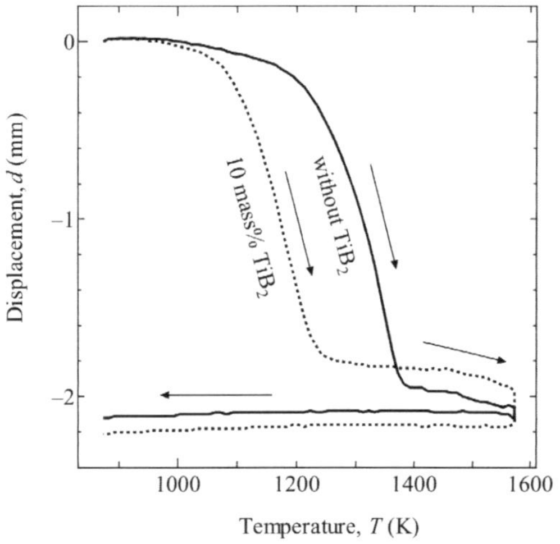

Figure 3. Densification behaviours of $BaTiO_3$ with and without 10 mass% TiB_2 during SPS.

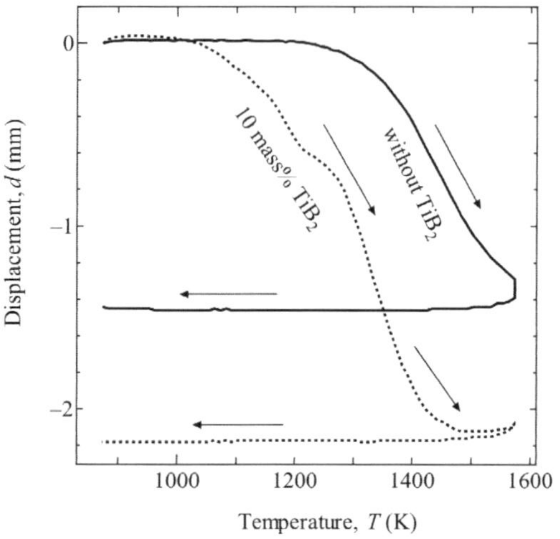

Figure 4. Densification behaviours of Y_2O_3 with and without 10 mass% TiB_2 during SPS.

temperature during sintering was markedly reduced by TiB_2 addition.

Figure 2 shows the XRD patterns of $SrTiO_3$ samples with and without 10 mass% TiB_2 after SPS process. The sample without TiB_2 was composed of the $SrTiO_3$ single phase, indicating the perovskite structure of the mother powder was maintained after the sintering. In the XRD pattern of the $SrTiO_3$ with TiB_2 addition, the peaks from TiB_2 phase were also detected and there was no other phase due to reaction of the $SrTiO_3$ and TiB_2 phase, resulting in the $SrTiO_3/TiB_2$ composite compact. From these results, it was found that the $SrTiO_3$ and TiB_2 phases were stably coexistent under the sintering condition of SPS in this study. As described above, the $SrTiO_3$ perovskite type oxide has a high melting point and its consolidation process to obtain a densely sintered body generally requires a sintering process at high temperature. As shown in figure 1, the TiB_2 powder addition is quite effective for reducing the densification temperature during SPS of the $SrTiO_3$ phase. Next, we investigated the possibility of the TiB_2 addition on consolidation at lower temperature of other oxides, such as $BaTiO_3$ and Y_2O_3. Figures 3 and 4 shows the shrinkage curves of $BaTiO_3$ and Y_2O_3, respectively, with and without 10 mass% TiB_2 during the SPS process. In the case of the $BaTiO_3$ system, the rapid densification of the sample without TiB_2 occurred around 1210 K and finished at about 1400 K. When 10 mass% TiB_2 was added, as shown in figure 3, the consolidation of powder sample proceeded at temperature range between 1060 K and 1260 K, which is lower by about 150 K than that of the sample without TiB_2. According to the XRD analysis, the sintered body was composed of $BaTiO_3$ and TiB_2 phases only, and formation of any other phase resulted from reaction between these mother powders was not observed. The results of XRD analysis and sintering behaviour when TiB_2 was added in the $BaTiO_3$ system were quite similar to those of the $SrTiO_3$ system. On the other hand, in the case of the Y_2O_3 without addition, even when sintering temperature reached 1573 K, the densification of the sample was not over. However, the compact shrinkage of the sample with TiB_2 addition occurred at lower temperature by about 250 K than the sample without addition and finished at about 1480 K. The effect of TiB_2 addition on reducing densification temperature was found to be superior to those in the $SrTiO_3$ and $BaTiO_3$ systems. Besides that, the final densification of the sample with TiB_2 after sintering was also quite large as compared to that of the Y_2O_3 sample without addition. The XRD analysis showed that there were small peaks from Y_3BO_6 phase detected in addition to the peaks from Y_2O_3 and TiB_2 phases in the pattern, indicating that the reaction between Y_2O_3 and TiB_2 powders partially occurred at the interface of these phases. This reaction at intergrain is considered to promote the particle rearrangement for densification and/or the disappearance of pores under the applied pressure during sintering, resulting in the large shrinkage.

4. Discussions

As shown in figures 1, 3 and 4, it was found that TiB_2 addition was quite effective for decreasing the densification temperature during SPS process of several oxides. Thus, it is expected that the sintering temperature required to obtain a densely sintered body can be significantly reduced by TiB_2 addition, as compared to the SPS process under the conventional condition (without TiB_2 addition). On the other hand, as shown in figure 2, after SPS of the powder mixture of $SrTiO_3$ and TiB_2, the sintered body was composed of $SrTiO_3$ and TiB_2 phases only, and formation of any other phase resulted from reaction between these mother powders was not observed. This experimental result implies that reactive sintering and/or liquid phase sintering, which can promote densification process, did not occur. Figure 5 shows the SEM photograph of the $SrTiO_3/TiB_2$ sample after SPS process. The black particles of TiB_2 phase were dispersed in the $SrTiO_3$ gray matrix, which is well consistent with the result shown in figure 2. There was no trace of third phase formation between these mother phases observed at the $SrTiO_3/TiB_2$ interface. Thus, it is confirmed that the sample formed a $SrTiO_3/TiB_2$ simple compact without any reacted phases even in the microscopic scale.

Generally, the reduction in densification temperature is caused by reactive sintering and/or liquid phase sintering. However, in the case of the $SrTiO_3/TiB_2$ simple composite, the consolidation at lower temperature observed in this study is not considered to be due to these sintering mechanisms. On the other hand, in the SPS process, the pulsed large current is directly applied to the powder compact. TiB_2 is well known as a good electrical conductor, and the electrical resistivity of the sample with TiB_2 addition is smaller than that of the sample without addition [6]. Therefore, it is considered that a larger amount of pulsed current is applied to the $SrTiO_3/TiB_2$ composite sample as compared to the sample without TiB_2 addition during heating, which may cause the reduction in densification temperature of the composite sample. Based on this consideration, $SrTiO_3$ powder was mixed with 10 mass% of Cu powder, which is also a good electrical conductor, instead of TiB_2 powder, and then sintered through the SPS process in the same condition, in order to investigate the effects of Cu addition on densification temperature. In figure 6, the shrinkage curve of the $SrTiO_3$/Cu composite during SPS process was compared with those of the $SrTiO_3$ with and without TiB_2 addition. It is obvious that the sample with Cu addition started shrinking at about 1110 K, which is almost the same as that of the $SrTiO_3$ without TiB_2 and then the densification finished around 1500 K, which is rather

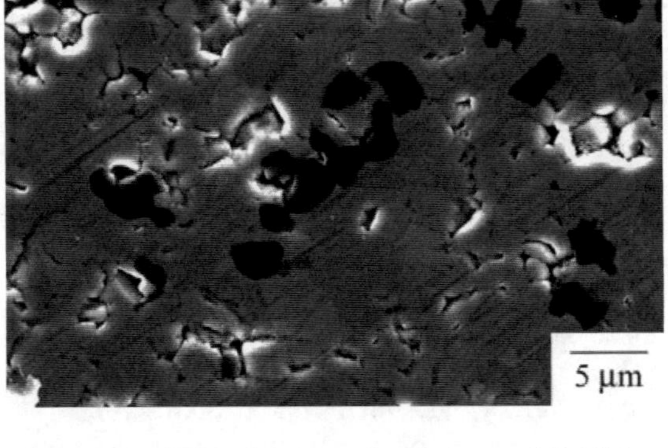

Figure 5. SEM photograph of sintered $SrTiO_3$ with 10 mass% TiB_2.

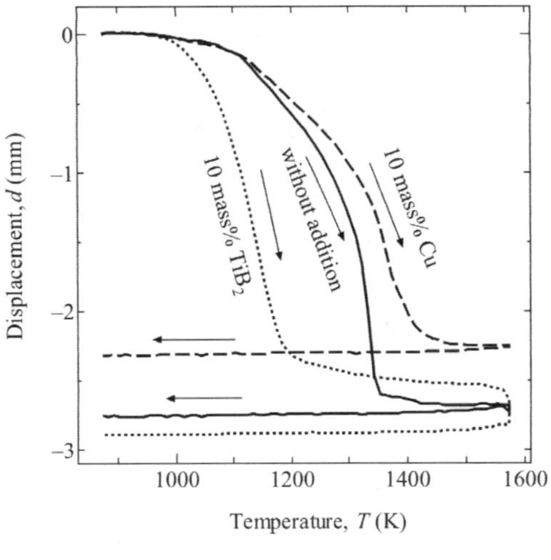

Figure 6. Densification behaviours of $SrTiO_3$ with and without 10 mass% TiB_2 and Cu during SPS.

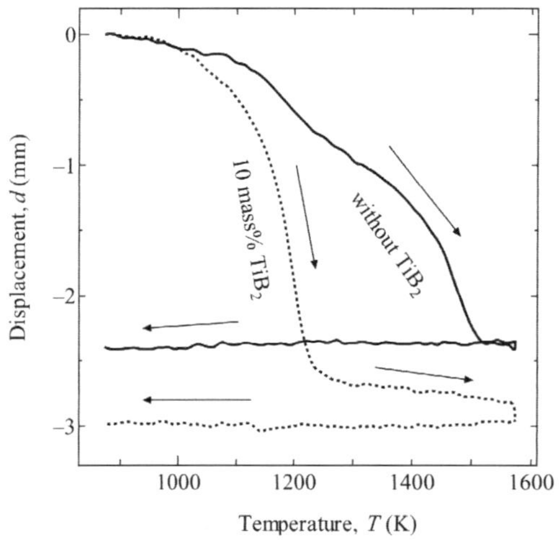

Figure 7. Densification behaviours of $SrTiO_3$ with and without 10 mass% TiB_2 during hot pressing.

higher than that of the $SrTiO_3$ without addition. This fact indicated that the effect of TiB_2 addition on decreasing in consolidation temperature was not caused by the reduction in electrical resistivity of the sample. From this experimental result, it is expected that the effect of TiB_2 addition obtained during the SPS process is also available when the sample is sintered by other sintering process. Generally, the Y-doped n-type thermoelectric $SrTiO_3$ was sintered by hot pressing at about 1673 K [3,7,8]. Then, the powder mixture of $SrTiO_3$ and TiB_2 was sintered by the conventional hot pressing and its sintering behaviour was observed. Figure 7 shows the shrinkage curves of the $SrTiO_3$ samples with and without 10 mass% TiB_2 during hot pressing. It was found that the densification temperature was significantly decreased by TiB_2 addition, and this sintering behaviour was well consistent with that when the sample was sintered by SPS process, indicating that the TiB_2 addition is also effective for reducing densification temperature in the case of the conventional sintering process. It is required that the mechanism causing the densification at low temperature by TiB_2 addition will be clarified in the near future by further investigation on microstructure at intergrain and/or formation of smaller amount of precipitates etc.

5. Conclusions

$SrTiO_3$ powder was mixed with TiB_2 powder and then the powder mixture was sintered by using the SPS. The TiB_2 addition significantly reduced the densification temperature of $SrTiO_3$ compact by about 150 K. After the SPS, the sintered body was composed of $SrTiO_3$ and TiB_2 phases only, and formation of any other phase resulted from reaction between these mother powders was not observed. This experimental result implies that reactive sintering and/or liquid phase sintering, which can promote densification process, did not occur. When $BaTiO_3$ powder and Y_2O_3 powder were sintered with TiB_2 powder under the same condition of SPS, the reduction in densification temperature was also observed. However, Cu powder addition, instead of TiB_2 powder, hardly affected the densification behaviour of $SrTiO_3$. The powder mixture of $SrTiO_3$ and TiB_2 was also sintered by the conventional hot pressing. The TiB_2 addition was also found to be quite effective for reduction of densification temperature even when sintered by the conventional sintering process. Thus, TiB_2 is considered to be a potential candidate as a sintering agent for low temperature densification of oxides.

References

[1] Moos R, Gnudi A and Hardtl K H, 1995 *J. Appl. Phys.* **78** 5042-47

[2] Hashimoto S, Kindermann L, Poulsen F W and Mogensen M, 2005 *J. Alloy. Compd.* **397** 245-49.

[3] Obara H, Yamamoto A, Lee C H, Kobayashi K, Matsumoto A and Funahashi R, 2004 *Jpn. J. Appl. Phys.* **43** L540-42

[4] Okuda T, Nakanishi K, Miyasaka S and Tokura Y, 2001 *Phys. Rev. B* **63** 113104.

[5] Ohta S, Nomura T, Ohta H, Hirano M, Hosono H and Koumoto K, 2005 *Appl. Phys. Lett.* **87** 092108.

[6] Ito M, 2009 *J. Jpn. Soc. Powder Powder Metall.* **47** 371-76 (in Japanese)

[7] Ito M and Ohira N, 2008 *Mater. Trans.* **49** 1844-47

[8] Ito M and Matsuda T, 2009 *J. Alloy. Compd.* **477** 473-77

Effects of Hydrogen Incorporation in the Formation of Hydrogenated Diamond-like Carbon Films

Yasuo Murakami[1], **Koji Hosaka**[2], **and Satoshi Hamaguchi**[1]
[1]Center for Atomic and Molecular Technologies, Osaka University, 2-1 Yamadaoka, Suita, Osaka 565-0871, Japan.
[2]Kansetsu corporation, 2-1-7 Shinkawa, Chuo-ku, Tokyo 104-0033, Japan

E-mail: murakami@ppl.eng.osaka-u.ac.jp

Abstract. The formation process of sp^3 hybridized carbon networks (i.e., diamond-like structures) in hydrogenated diamond-like carbon (DLC) films has been studied with the use of molecular dynamics (MD) simulations. A recent study [Y. Murakami, S. Horiguchi, and S. Hamaguchi, Phys. Rev. E. (2010) *in press*] has shown that a sufficient supply of hydrogen in the form of hydrocarbon species CH_3 as deposition precursors significantly enhances the formation of sp^3 hybridized bond structures in the deposited film. In this study, we have examined the role of hydrogen in such deposition processes by simulating the cases where carbon and hydrogen atoms are supplied to the surface separately. It has been found that, in such a case, the rate of sp^3 bond formation is significantly lower than that in the corresponding case of CH_3 injections. The result suggests that energetic hydrogen atoms present in the vicinity of the incident carbon atoms promote the formation of σ bonds (and therefore sp^3 bonds) among carbon atoms by increasing the rates of π bond termination and hydrogen abstraction reactions.

1. Introduction

Amorphous carbon with a high content of sp^3 structures is often called diamond-like carbon (DLC) and known to have extraordinary material properties such as high mechanical hardness, high wear resistance, low friction coefficients, high chemical inertness, good infrared (IR) transparency, high electrical resistivity, good field emission properties, and low dielectric coefficients [1-7]. For example, DLC is used widely as protection layers for data storage disks and magnetic recording heads.

Characteristics of DLC films are generally determined by the amount of sp^3 hybridized bonds present in the films [2]. In the present study, we examine sp^3 hybridized bond formation mechanisms in hydrogenated amorphous carbon (a-C:H) deposition processes. In recent studies [8,9] based on numerical simulations, it has been shown that the presence of a sufficient amount of hydrogen in an amorphous carbon film being deposited significantly enhances the formation of sp^3 hybridized bond structures if the incident (hydro)carbon beam has a sufficiently high kinetic energy. In these studies, hydrogen was supplied to the surface in the form of hydrocarbon species CH_x as deposition precursors. In the present study, we simulate the cases where carbon and hydrogen atoms are supplied to the surface separately in order to clarify the role of hydrogen in such deposition processes.

As a tool for the study, we use classical molecular dynamics (MD) simulation, by which dynamics of individual atoms may be examined in detail. In this work, we examine the initial phase of carbon film deposition by numerically simulating injection processes of carbon, hydrogen, or hydrocarbon atoms/clusters into amorphous carbon (a-C) films that are initially hydrogen-free. These results may be compared to the results given in Ref. [9], where injection processes of CH and CH_3 were examined. It is of our special interest here whether hydrogen injected separately from the incident carbon atoms can also enhances the formation of sp^3 structures in a carbon deposition process.

The rest of the paper is arranged in the following manner. We first briefly present our MD simulation scheme in the next section. In Sec. 3, simulation results are presented and discussed. The conclusions are given in the final section.

2. Molecular Dynamics Simulation

In classical MD simulations, trajectories of all atoms in the system are obtained from the integration of Newton's equations of motion [10-13] and macroscopic physical quantities, such as deposition rates and sputtering yields, may be obtained from proper averaging of corresponding microscopic quantities. Atomic interactions in classical MD simulations are predetermined and the interatomic potential functions used in this study are multi-body potential functions that are the same as those used in Refs. [8] and [9]. More details on the functions may be found in Ref. [11].

The model substrate used in the simulation is a hydrogen-free amorphous carbon (a-C) cube with an edge length of about 2 nm. It consists of approximately 800 carbon atoms that were initially arranged at random positions in the cube and then brought to thermal equilibrium at 300 K by MD simulations of canonical ensemble (with constant temperature and density). Its mass density is about 2.0 g/cm^3, which is typical for a-C. The model substrate consists mainly of sp^2 hybridized bonds; the ratio of the number of sp^3 carbon atoms to that of sp^2 atoms in the model substrate is about 0.4%.

The periodic boundary conditions are imposed in the lateral directions and therefore the model substrate represents an infinitely wide flat thin film with a thickness of about 2 nm. Atoms in the 0.4 nm thick bottom layer of the substrate are set immobile lest the substrate drift downward when its top surface is subject to beam injections.

In this work, the beam species that we have chosen to examine are C, H, and CH_3. It has been known that CH_3 is the most abundant radical species in methane plasmas that are often used for a-C:H deposition in Plasma-Enhance Chemical Vapour Deposition (PECVD) processes [14-17]. Methyl ion CH_3^+ is known to form one of the largest ion fluxes emitted from methane plasmas [15]. Energetic beams simulated in this work are considered to represent ion beams. However, all interatomic forces used in the simulations are those for charge neutral species for the sake of simplicity.

The sequence of deposition simulations is the following [8-13]. A single simulation cycle starts with an injection of the selected species (such as CH_3) at a predetermined kinetic energy into the substrate surface. The substrate is set in thermal equilibrium at 300 K. The initial lateral position of the injected species is randomly selected and its initial height is set to be about 1 nm above the substrate surface. The incident angle is normal to the substrate surface. The trajectories of all atoms are then obtained from the integration of Newton's equations of motion under the condition of constant total energy, i.e., microcanonical conditions, for 0.8 ps. The numerical time integration is based on the velocity Verlet algorithm. The system is then rapidly cooled and equilibrated at the initial temperature (300 K) during the following 0.2 ps to remove the excess heat from the system. The next simulation cycle then starts with a new particle injection, as in the first cycle. In the MD simulation study presented here, such simulations cycles are typically repeated 500-800 times, which corresponds to the beam dose (i.e., fluence) of $1.25\times10^{16} \sim 2.0\times10^{16}$ cm^{-2}.

3. Simulation results

In this work, we performed four different sets of simulations in order to clarify the formation mechanisms of sp^3 bonds in the deposited carbon films. Hardness and other industrially useful characteristics of a DLC film is determined by the sp^3 carbon content in the film. Table 1 shows the four different simulation conditions employed in this work.

In the simulation case denoted by Condition I, carbon atoms are injected 800 times into the model substrate. As discussed above, in each injection cycle, the MD simulation is run under the microcanonical condition for 0.8 ps after a single carbon atom C is released from a location above the model substrate surface with a kinetic energy of 50 eV in the direction normal to the substrate surface. The system is then cooled for 0.2 ps to 300 K. Similarly, in the case denoted by Condition II, CH_3 (instead of C) atomic clusters are injected into the model substrate surface with the same conditions.

Table 1. Conditions for simulations presented in this article.

Conditions	I	II	III		IV
Species	C	CH_3	C	H	H
Number per injection cycle	1	1	1	3	200
kinetic energy (eV) per species	50	50	40	3.3	0.025
Number of Injection cycles	800	800	660		500

In Condition III, the single simulation cycle consists of two separate sets of injections. In the first set of injection, as in Condition I, a single C atom is injected with a kinetic energy of 40 eV. The system is under the microcanonical condition for 0.8 ps and then cooled for 0.2 ps to 300 K, as in the Conditions I and II. Then in the second set of injections, three H atoms are injected simultaneously from random locations above the substrate surface with a kinetic energy of 3.3 eV for each. MD simulations are run under the microcanonical conditions for 0.8 ps and then cooled for 0.2 ps to 300 K, as in the first set. In Condition III, the total kinetic energy of C and three H atoms is 50 eV, which is the same kinetic energy of CH_3 of Condition II. The main difference between Condition II and Condition III is that, in Condition III, injected hydrogen atoms may not be in the vicinity of the injected C atom as their locations are selected randomly.

In Condition IV, 200 H atoms are supplied simultaneously to the substrate surface with sufficient low energy in a single injection cycle. The other conditions are similar to those in Conditions I and II.

Figure 1 shows the fully carbon terminated sp^3 fractions as functions of injection dose of C atoms under the Conditions I (denoted by the thin solid curve), II (broken curve), and III (thick solid curve). Here the "fully carbon terminated sp^3 fraction" means the fraction of fully carbon terminated sp^3 C atoms (each of which is bonded with 4 other C atoms only, not more and not less) for all deposited C atoms. This fraction represents the "diamond-likeness" of the deposited film. It should be noted that when the dose is small (i.e., the number of all deposited C atoms is small), the fractions given in Fig. 1 is statistically meaningless. However Fig. 1 shows that there are clear differences among the three cases when the dose is sufficiently large. The case of Condition II is the same as that shown in Fig. 6 of Ref. [9].

It is clearly seen in Fig. 1 that fully carbon terminated sp^3 structures are more efficiently formed under Condition II or III than Condition I, in which no hydrogen is provided to the surface. It indicates the importance of the role of hydrogen for efficient formation of diamond-like structures in carbon deposition processes.

Figure 1 also shows that a relatively large sp^3 fraction is obtained as the injection does of CH_3 increases under Condition II whereas the sp^3 fraction remains relatively small under Condition III, in which the same number of C and H atoms with the same kinetic energy for each atom are supplied to the substrate surface. Since a 50 eV CH_3 atomic cluster tends to break up when it hits the surface, one might expect that a similar process would occur after the injected C atom hits the substrate surface in both Conditions II and III. The major difference between these two cases is that, in Condition II, broken up H atoms may remain still in the vicinity of the injected C atom and therefore contribute to

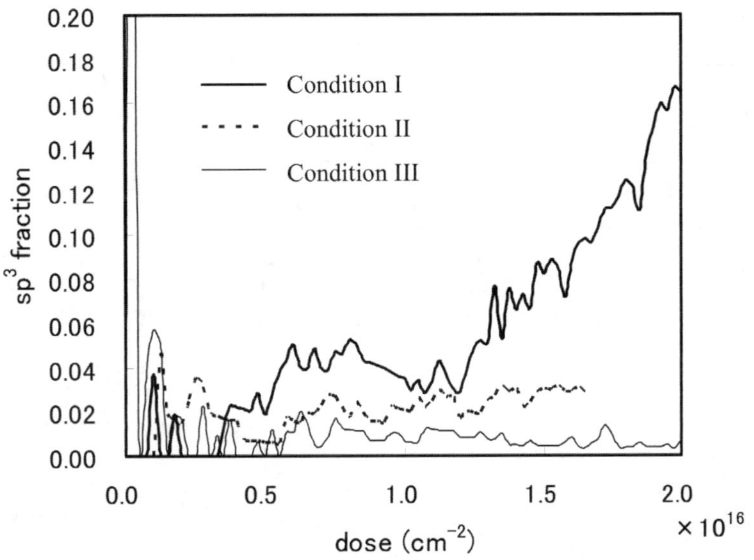

Figure 1. The fully carbon terminated sp^3 fractions as functions of the carbon dose for Conditions I (denoted by thin solid curve), II (broken curve), and III (thick solid curve) given in Table 1.

the formation of C-to-C σ bonds by bond scissions and/or hydrogen abstraction due to their relatively high kinetic energies.

For Condition IV, on the other hand, very few sp^3 structures have been observed to be formed. Therefore the sp^3 fraction under Condition IV is not given in Fig. 1.

Figure 2 shows depth profiles of the (a) C-to-C σ bond density and (b) carbon sp^2 bond density at various beam carbon doses under Condition II, obtained from MD simulations. Here the sp^2 bond means a carbon bond that contains a (partial) π orbital, such as the carbon bond of graphite. The horizontal axis represents the film thickness measured from the bottom of the model substrate. Each curve represents the density profile at the specified carbon dose. It is seen that, under Condition II, the C-to-C σ bond increases rapidly in the region below the top surface whereas the carbon sp^2 bond density in the same region gradually decreases. As seen in Fig. 1, the sp^3 fraction starts to increase at around 1.2×10^{16} cm^{-3} of carbon dose. Around this dose, the rapid decrease of carbon sp^2 bonds in the region also occurs, as indicated in Fig. 2(b). The decrease of sp^2 bond structures indicates that hydrogen atoms are sufficiently supplied to the substrate to the extent that a large number of π bonds there have been terminated by the supplied hydrogen atoms. Therefore we conjecture that, once the surface and subsurface regions are nearly saturated with hydrogen, energetically injected C atoms have higher chances of converting two carbon-to-hydrogen (C-H) bonds to a single carbon-to-carbon (C-C) σ bond in the region by promoting hydrogen abstraction reactions.

To compare the sp^3 formation processes between Conditions II and III, we present Fig. 3, where the depth profiles of the (a) C-to-C σ bond density and (b) carbon sp^2 bond density at various beam carbon doses under Condition III. Although the σ bond density increases near the surface region, as in the case of Fig. 2 (a), the sp^2 bond density also increases in the same region. This is the trend opposite to that seen in Fig. 2(b), in which the sp^2 bond density decreases as the injection dose increases. At this moment, it is not clear why the sp^2 bond density increases under Condition III. It is possible that hydrogen atoms injected at a few eV may convert freshly formed sp^3 structures to sp^2 structures. However, abundant hydrogen supply tends to break π bonds and therefore could reduce the number of sp^2 structures, as we have seen in Fig. 1.

Figure 4 also shows the depth profiles of the (a) C-to-C σ bond density and (b) carbon sp^2 bond density at various hydrogen doses under Condition IV, where low-energy H atoms are supplied to the surface without carbon injections. It is clearly seen that sp^2 bond structures of the initial substrate are

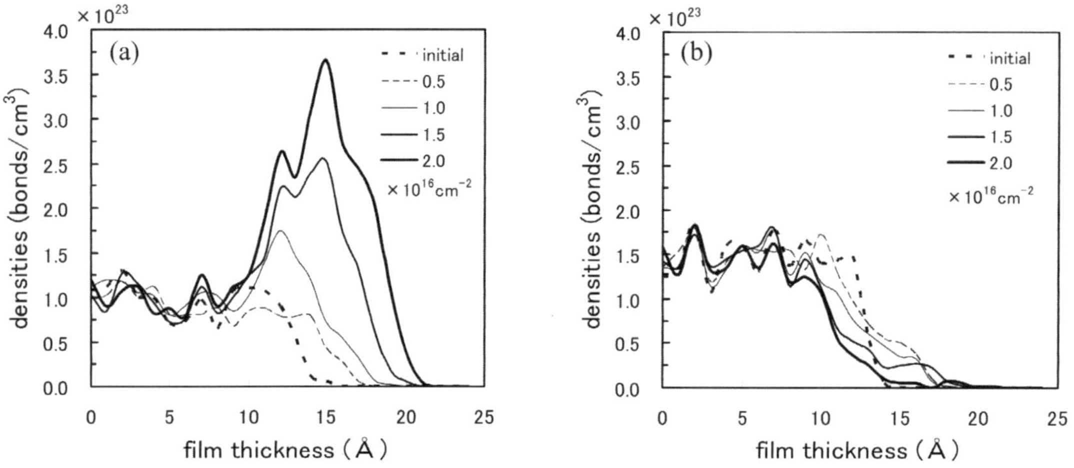

Figure 2. The depth profiles of the (a) C-to-C σ bond density and (b) carbon sp^2 bond density at various beam doses in Condition II, i.e., the case of 50 eV CH_3 beam incidence, obtained from MD simulations. The horizontal axis represents the film thickness measured from the bottom of the model substrate [9].

reduced in this process, as one expects. It is also seen that the C-to-C σ bond (i.e., single bond) density increases near the surface, which is a mere consequence of conversion of covalent bonds with π orbitals to pure C-to-C σ bonds due to breakup of the π bonds by supplied hydrogen.

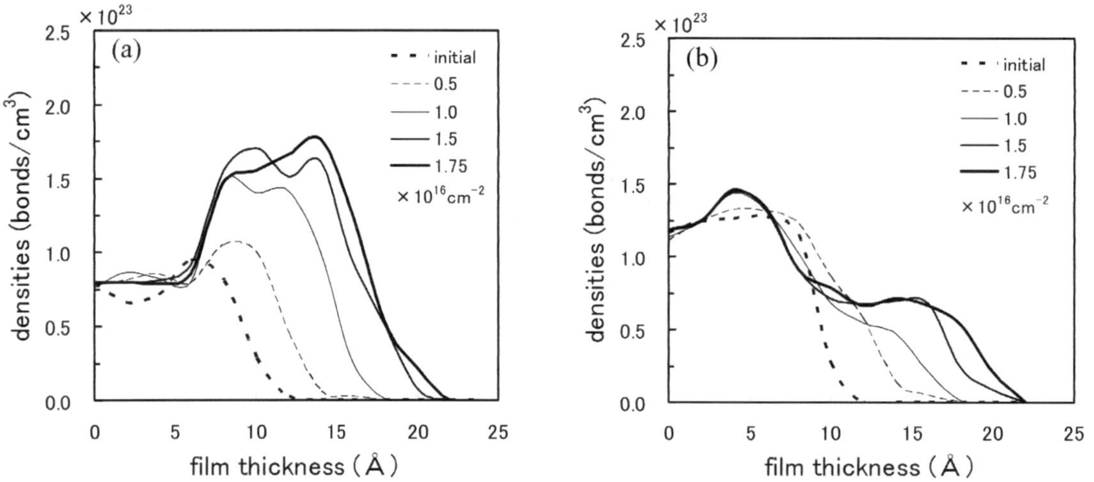

Figure 3. The depth profiles of the (a) C-to-C σ bond density and (b) carbon sp^2 bond density at various beam doses in Condition III, i.e., the separate injections of C at 40 eV and three H atoms at 3.3 eV each, obtained from MD simulations.

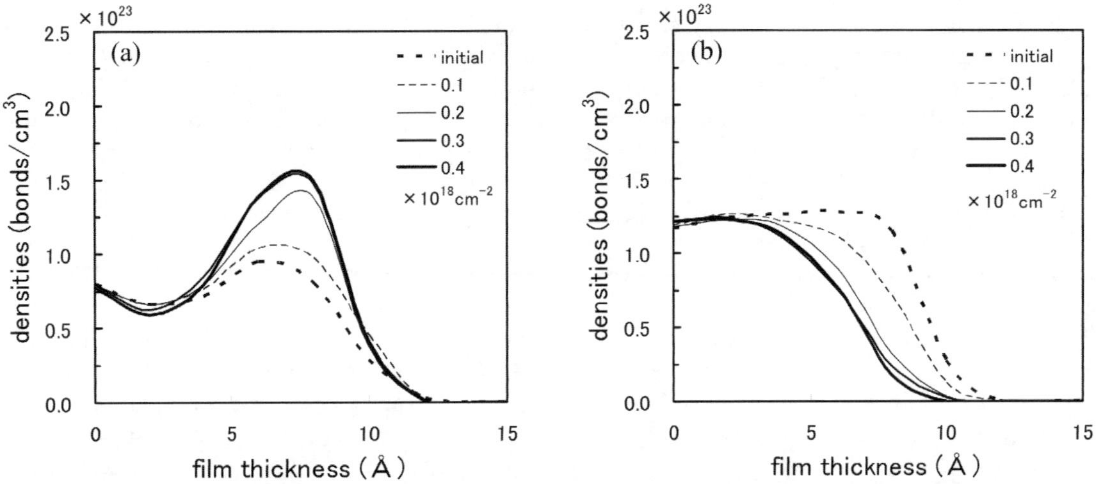

Figure 4. The depth profiles of the (a) C-to-C σ bond density and (b) carbon sp^2 bond density at various beam doses in Condition IV, i.e., injections of H atoms at low energy, obtained from MD simulations.

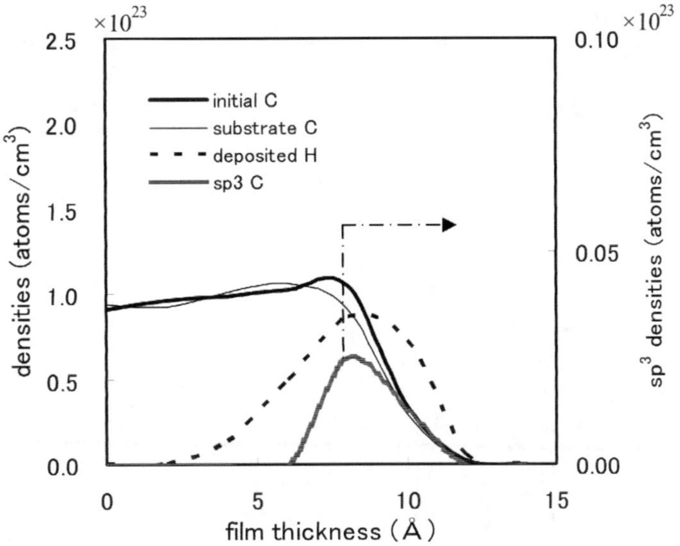

Figure 5. The depth profile of fully carbon terminated sp^3 C atoms at a dose of 4.0×10^{16} cm^{-2} under Condition IV (depicted by the grey curve denoted by "sp^3 C" and its value is given on the right scale), obtained from MD simulations. The profiles of C atoms including sp^3 C atoms in the initial (thick solid curve) and final (thin solid curve) states as well as the profile of deposited H atoms (dotted curve) are also plotted. Their values are given on the left scale. It is clearly seen that the number density of sp^3 C atoms formed in this process is significantly smaller than the total carbon density of the substrate.

Even under Condition IV, a very small number of fully carbon terminated sp^3 bonds are formed. Shown in Fig. 5 are the depth profile of fully carbon terminated sp^3 C atoms at a dose of 4.0×10^{16} cm^{-2} under Condition IV (depicted by the grey curve denoted by "sp^3 C" and its value is given on the right scale), the profiles of C atoms including sp^3 C atoms in the initial (thick solid curve) and final (thin solid curve) states, and the profile of deposited H atoms (dotted curve). The values for the C and H densities are given on the left scale. It is clearly seen that the number density of sp^3 C atoms formed in this process is significantly smaller than the total carbon density of the substrate.

4. Concluding remarks

We have examined the formation processes of fully carbon terminated sp^3 bonds under four different conditions summarized in Table 1, using MD simulations. It has been observed in the simulations that the presence of hydrogen in carbon deposition processes promotes the formation of sp^3 bonds in general. However, injections of carbon in the form of CH_3, rather than C and H atoms separately under similar conditions, exhibit the sp^3 formation at much higher rates. Although the exact cause for the enhanced rate of sp^3 bond formation has not been clear yet, the observations suggest that the presence of hydrogen with a kinetic energy of a few eV in the vicinity of the incident C atoms significantly promote the formation of the C-to-C σ bonds by bond scissions and/or hydrogen abstraction. More detailed examination of the role of energetic hydrogen randomly injected into a-C films in the sp^3 formation processes is deferred to a future study.

Acknowledgments

The work is partially supported by a Grant-in-Aid for Scientific Research from Ministry of Education, Culture, Sports, Science and Technology (MEXT) Japan. The authors are also grateful for the technical support for computer simulations by Mr. M. Isobe of Osaka University and Dr. M. Yamashiro of Nihon University.

References

[1] Robertson J 1992 *Surf. Coating Technol.* **50** 185.
[2] Jacob W and Möller M 1993 *Appl. Phys. Lett.* **63** 1771.
[3] Robertson J 2002 *Mater. Sci. Eng.* R**37** 129.
[4] Lifshitz Y 1996 *Diamond Relat. Mater.* **5** 388.
[5] Robertson J 1997 *Phil. Mag. B* **76** 335.
[6] Grill A 1999 *Diamond Relat. Mater.* **8** 428.
[7] Casiraghi C, Ferrari A C, and Robertson J 2005 *Phys. Rev. B* **72** 085401.
[8] Murakami Y and Hamaguchi S 2009 *J. Plasma Fusion Res.* **85** 674 (*in Japanese*).
[9] Murakami Y, Horiguchi S, and Hamaguchi S 2010 *Phys. Rev. E. in press.*
[10] Yamada H and Hamaguchi S 2004 *J. Appl. Phys.* **96** 6147.
[11] Yamada H and Hamaguchi S 2005 *Plasma Phys. Control. Fusio*n **47** A11.
[12] Yamashiro M, Yamada H, and Hamaguchi S 2007 *J. Appl. Phys.* **101** 046108.
[13] Yamashiro M, Yamada H, and Hamaguchi S 2007 *Jpn. J. Appl. Phys.* **46** 1692.
[14] Mutsukura N, Inoue S, and Machi Y 1992 *J. Appl. Phys.* **72** 43.
[15] Kleber R, Weiler M, Krüger S, Sattel S, Kunz G, Jung K, and Ehrhards H 1993 *Diamond Relat. Mater.* **93** 246.
[16] Pecher P and Jacob W 1998 *Appl. Phys. Lett.* **73** 31.
[17] Garrison B J, Kodali P B S, and Srivastava D 1996 *Chem. Rev.* **96** 1327.

Effect of Nb and Zr additions to magnetic properties of Nd-Fe-B bulk nanocomposites

R Kawai[1], T Kamoi[1], T Fukuzaki[2], R Tamura[1,2], K Nishio[1], A Hayashi[3], N Aikawa[3]

[1] Department of Materials Science and Technology, Tokyo University of Science, Noda, Chiba, 278-8510, Japan

[2] Polyscale Technology Research Center, Tokyo University of Science, Noda, Chiba, 278-8510, Japan

[3]Department of Applied Electronics, Tokyo University of Science, Noda, Chiba, 278-8510, Japan

E-mail:hitonatumanatu@gmail.com

Abstract. We have investigated the microstructure and the coercivity of $Nd_{10}Fe_{77.5}B_{12.5-x}M_x$ (M=Zr, Nb, x=0~3) bulk nanocomposite magnets in order to obtain insight into the factors responsible for the coercivity increase by the addition of Zr and Nb. The volume fraction of the $Nd_2Fe_{14}B$ phase is found to increase appreciably with the Zr and Nb addition, indicating that the Zr/Nb addition is effective in the suppression of a second phase. The increase of the coercivity in both alloys is understood to be due to the increase of the volume fraction of the $Nd_2Fe_{14}B$ hard phase.

1. Introduction

Nd-Fe-B nanocomposite magnets exhibit high magnetic performance despite for their reduced Nd concentrations from the stoichiometric composition, i.e., $Nd_2Fe_{14}B$, and have been investigated extensively for future applications [1]. However, the coercivities of ternary Nd-Fe-B nanocomposite bulk magnets prepared by the spark plasma sintering method are not high enough and an improvement of the coercivity is required in order to achieve a high maximum energy product. A number of factors are known to influence the coercivity of nanocomposite magnets such as the grain size, the volume fraction of the $Nd_2Fe_{14}B$ hard phase and the condition of the grain boundaries[2-6]. It has been reported that an addition of Zr and Nb is effective in the improvement of the coercivity of Nd-Fe-B nanocomposite magnets [7]. Following the previous works, we have studied the microstructure and the coercivity of $Nd_{10}Fe_{77.5}B_{12.5-x}M_x$ (M=Zr, Nb, x=0~3) alloys in order to obtain insight into the key factor responsible for the coercivity increase induced by the addition of Zr and Nb.

2. Experimental

Alloys of nominal compositions $Nd_{10}Fe_{77.5}B_{12.5-x}M_x$ (M=Zr, Nb, x=0~3) prepared in an arc furnace were melt-spun with the surface velocity of 50 m/s of the Cu wheel. Melt-spun ribbons were then crushed into fine powders of < 100 μ m and consolidated by the spark plasma sintering (SPS) method under a pressure of 100 MPa during heating to 650 deg in 5 min and furnace-cooling to room temperature. Characterization of the samples was performed by X-ray diffraction (XRD) and differential scanning calorimetry (DSC). The microstructures of the samples were investigated by field emission scanning electron microscopy (FE-SEM). Magnetic properties were measured by using a vibrating sample magnetometer (VSM).

3. Results and discussion

Figure 1 shows X-ray diffraction patterns of $Nd_{10}Fe_{77.5}B_{12.5-x}M_x$ (M=Zr or Nb, x=0~3) alloys. It is clear that all the samples are composed mostly of the $Nd_2Fe_{14}B$ phase. Since no crystalline peak is observed in the as-spun samples, the $Nd_2Fe_{14}B$ hard phase is formed during the SPS sintering process. It is also seen that the peaks become sharper as the Zr or Nb concentration is raised. The sharpening of the peaks may suggest an improvement of the structural order of the $Nd_2Fe_{14}B$ phase by the Zr or Nb addition. Figure 2 shows DSC curves of $Nd_{10}Fe_{77.5}B_{12.5-x}M_x$ (M=Zr or Nb, x=0~3) as-spun ribbons. Two exthothermic peaks are clearly noticed in all the samples, indicating the precipitation of a second phase beside the $Nd_2Fe_{14}B$ phase. With increasing the Zr or Nb concentration, both the crystallization temperatures are observed to be shifted to higher temperatures, indicating that the stability of the amorphous state becomes higher with the additions of Zr and Nb, as been reported in the literatures[4][7]. Both the XRD pattern and the DSC curve for 3 at%Zr show that the second phase almost disappears by the addition of 3 at%Zr.

Figure 3 shows back scattering images of the $Nd_{10}Fe_{77.5}B_{12.5-x}M_x$ (M=Zr, Nb, x=0~3) alloys. In the figure 3, the bright regions represent the $Nd_2Fe_{14}B$ phase and the dark regions the second phase. It is seen from the figure, the area of the second phase reduces with the Zr/Nb addition. We have estimated the ratio of the bright and the dark areas in figure 3 by counting their pixels using an image-analysis software (GUIfigure) and the results are presented as a function of the Zr/Nb concentration in the figure 5. It is clear that the volume fraction of the second phase decreases with increasing the Zr/Nb concentration, indicating that the Zr/Nb addition is effective in the suppression of precipitation of the second phase from the amorphous matrix.

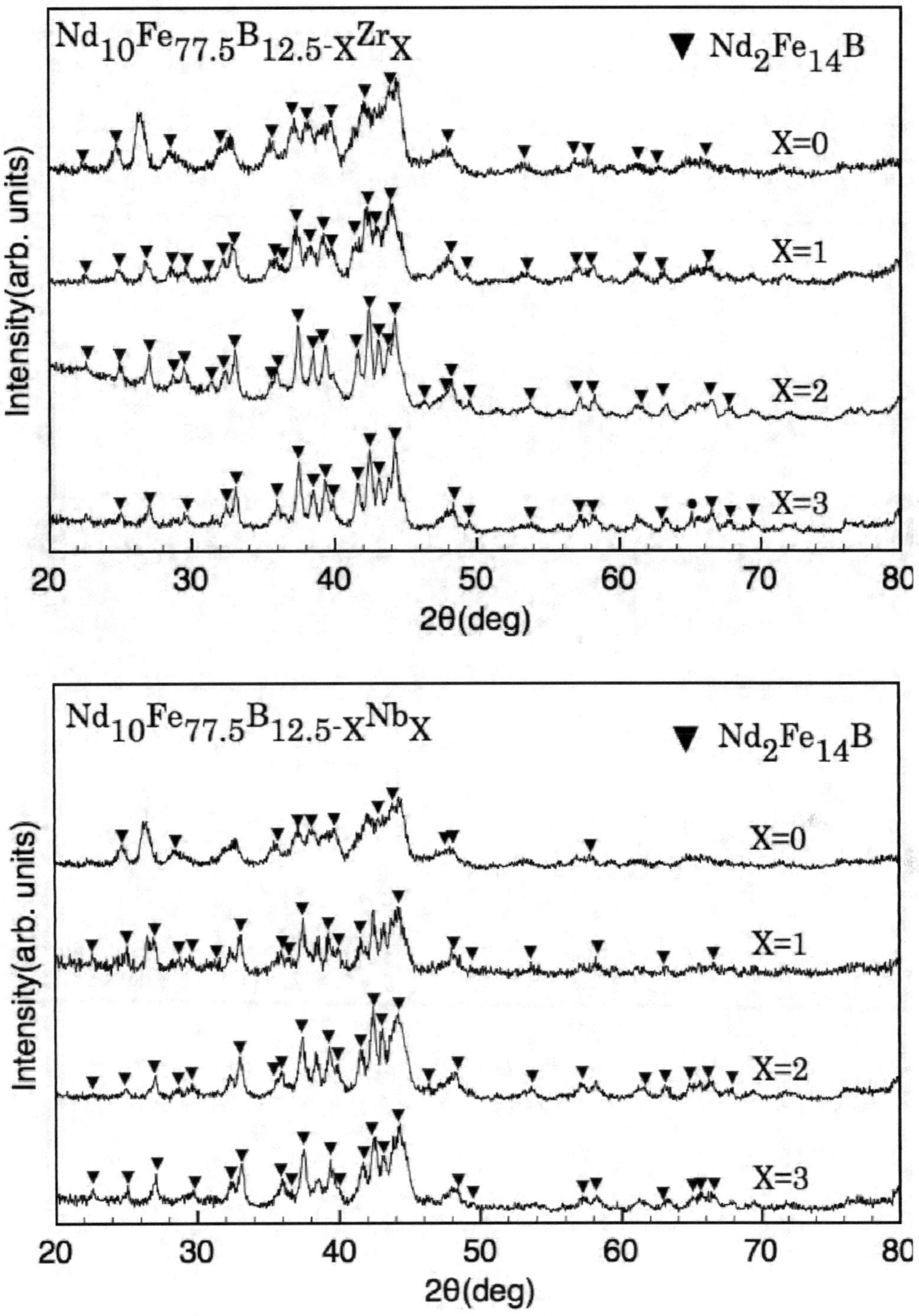

Figure 1. XRD patterns of $Nd_{10}Fe_{77.5}B_{12.5-x}M_x$ (M=Zr, Nb, x=0~3) bulk magnets prepared by SPS sintering

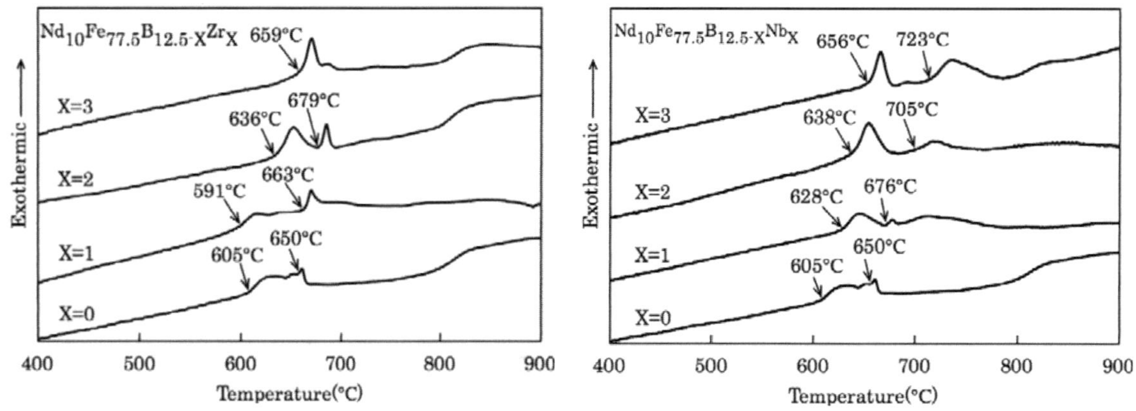

Figure 2. DSC curves of $Nd_{10}Fe_{77.5}B_{12.5-x}M_x$ (M=Zr, Nb, x=0~3) as-spun ribbons

$Nd_{10}Fe_{77.5}B_{12.5-X}Zr_X$ (X=0~3)

X=0 X=1 X=2 X=3

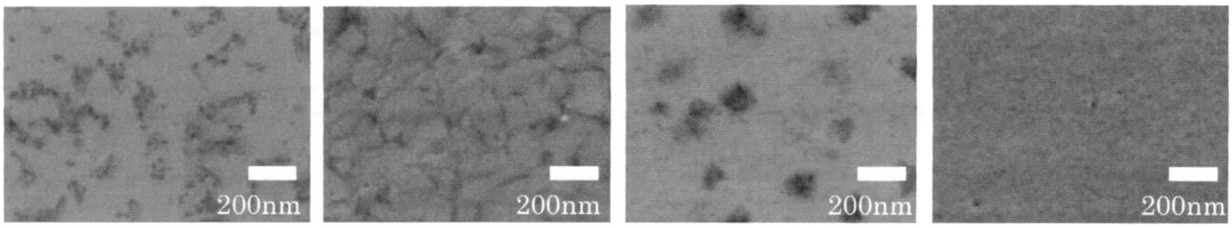

X=0 X=1 X=2 X=3

$Nd_{10}Fe_{77.5}B_{12.5-X}Nb_X$ (X=0~3)

Figure 3. Back scattering images of $Nd_{10}Fe_{77.5}B_{12.5-x}M_x$ (M=Zr, Nb, x=0~3) bulk magnets prepared by SPS sintering.

Figure 4 shows demagnetization curves of the $Nd_{10}Fe_{77.5}B_{12.5-x}M_x$ (M=Zr, Nb, x=0~3) alloys. The coercivity increases as the Zr/Nb concentration is raised, which is interpreted as a consequence of the increase of the volume fraction of the $Nd_2Fe_{14}B$ phase. The higher coercivity for 3 at%Zr than for 3 at%Nb is also understood by higher volume fraction of the $Nd_2Fe_{14}B$ phase in the Zr-added alloy.

Thus, the Zr addition is more effective in the suppression of the second phase and hence in the improvement of the coercivity.

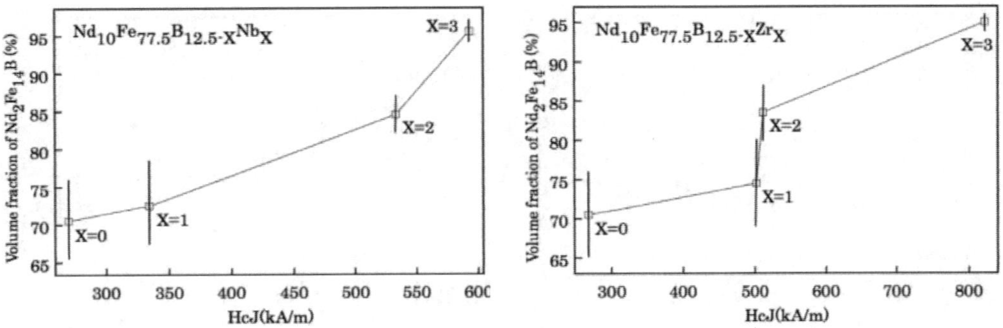

Figure 4. Volume fraction of the $Nd_2Fe_{14}B$ phase and the coercivity of the $Nd_{10}Fe_{77.5}B_{12.5-x}M_x$ (M=Zr, Nb, x=0, 3) bulk magnets plotted against the coercivity

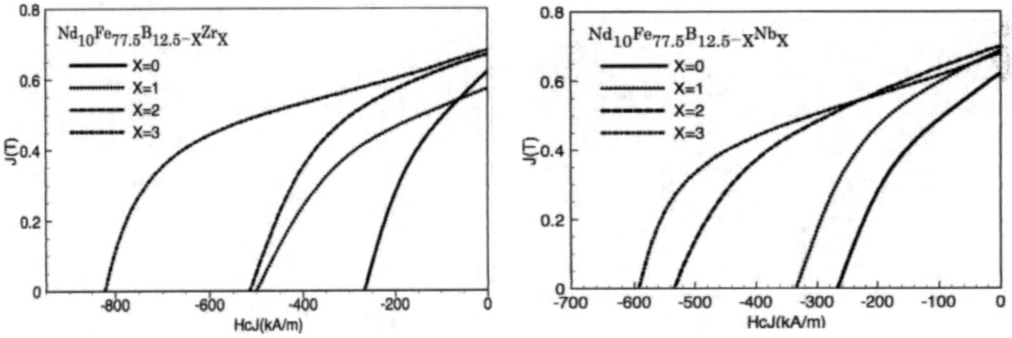

Figure 5. Demagnetization curves of the $Nd_{10}Fe_{77.5}B_{12.5\text{-}x}M_x$(M=Nb,Zr x=0~3) bulk magnets prepared by SPS sintering

4. Conclusion

We have investigated the effect of Zr and Nb additions to the microstructures and the magnetic properties of $Nd_{10}Fe_{77.5}B_{12.5-x}M_x$ (M=Zr, Nb, x=0~3) alloys prepared by the spark plasma sintering method. The volume fraction of the $Nd_2Fe_{14}B$ phase is found to increase with the Zr or Nb addition, indicating that the Zr/Nb addition is effective in the suppression of the precipitation of a second phase. The increase of the coercivity in both alloys is understood by the increase of the volume fraction of the $Nd_2Fe_{14}B$ hard phase.

Acknowledgments

This work was supported by "Academic Frontier" Project for Private Universities: Matching Fund Subsidy from MEXT (Ministry of Education, Culture, Sports, Science and Technology), 2006-2010.

A part of this study was conducted at the Center for Nano Lithography & Analysis, The University of Tokyo, and was supported by MEXT, Japan.

References

[1] Coehoorn R, De Mooij D B, and Waard C De *J. Mag. Mag. Mater*. **80**(1989), 101

[2] Hirosawa S, Kanekiyo H, Miyoshi T, *J. Mag. Mag. Mater*. **281**(2004), 58-67

[3] Wu Y Q, Ping D H, Xiong X Y, Hono K, *J. Appl. Phys*. **91**, 8174, 2002

[4] Ping D H, Hono K, Kanekiyo H and Hirosawa S, *Acta Mater*. **47**, 4641(1999)

[5] Hirosawa S, *Metal*, **69**, 692(1999)

[6] Hirosawa S, Kanekiyo H, Uehara M, *J. Apply. Phys*. **73**, 6488(1993)

[7] Hono K, Nikkan newspaper company of industry,(2005)

Synthesis of Fe-based nanocrystalline soft magnets with high saturation magnetization

T Kumakawa[1], T Fukuzaki[2], K Nishimoto[1] and R Tamura[1,2]

[1]Department of Materials Science and Technology, Tokyo University of Science, Noda, Chiba 278-8510, Japan

[2]Polyscale Technology Research Center, Tokyo University of Science, Noda, Chiba 278-8510, Japan

E-mail: j8209623@ed.noda.tus.ac.jp

Abstract. Fe-metalloid-based nanocrystalline soft magnets are known to be excellent soft magnetic materials. In this work, we have synthesized $(Fe,Co)_{87}(P_{7-x}C_1B_x)_{13}$ alloys with a high saturation magnetization (Bs) and investigated their magnetic properties and microstructures. As a result, we have successfully obtained a high saturation magnetization of Bs~1.85T at the compositions of $(Fe,Co)_{87}(P_1C_1B_6)_{13}$ and $(Fe,Co)_{87}(C_1B_7)_{13}$. The B addition is found to increase the saturation magnetization through increasing the volume fraction of nanocrystalline α-Fe whereas simultaneous additions of B and P are found to be effective in the grain refinement, resulting in low coercivity.

1. Introduction

Nanocrystalline soft magnetic materials have been developed in Fe-metalloid (P,C,B,Si) and Fe-metalloid-M(M=Ga,Zr,Nb,etc) systems[1-6]. For the improvement of nanocrystalline soft magnets, high (Fe, Co) concentration is required in order to achieve a high Bs. However, the increase of the (Fe, Co) concentration results in a lower glass forming ability (GFA), as a result, leading to an unfavorable increase of the coercivity[7-9]. It is known that the addition of a transition metal element M is effective for the improvement of the GFA. However, such an addition of M reduces the Bs, which is unfavorable. Therefore, in this work, for the purpose of developing high Bs nanocrystalline soft materials, we have performed optimization of the concentrations of semi-metallic elements such as B and P in order to improve the GFA while keeping the (Fe,Co) concentration at a substantially high level.

2. Experimental

Alloys of nominal compositions $(Fe,Co)_{87}(P_{7-x}C_1B_x)_{13}$(x=0,3,6,7) were prepared by induction melting of high purity elements Fe (99.9%), Co(99.9%), Fe_3P(99.9%) together with Fe-5.88wt%C and Fe-14.46wt%B alloys prepared by arc-melting. Then, the alloys were melt-spun into ribbons with the surface velocity of 50 m/s of the Cu wheel. Characterization of the samples was performed by X-ray diffraction (XRD) and the nanostructure was probed by transmission electron microscopy (TEM) observations. Magnetization measurements were performed by using a vibrating sample magnetometer (VSM) and the coercivity was measured by using a Hc meter.

3. Results and Discussion

Figure 1 shows X-ray diffraction patterns of $(Fe,Co)_{87}(P_{7-x}C_1B_x)_{13}$(x=0,3,6,7) melt-spun alloys. For x=0, the sample is mostly crystallized into α-Fe phase and for x=3,6, halo patterns are observed, which indicates the presence of nanocrystalline or amorphous phase. For x=7, an α-Fe peak is recognized suggesting an occurrence of substantial grain growth of α-Fe. Figure 2 shows TEM bright field images of $(Fe,Co)_{87}(P_{7-x}C_1B_x)_{13}$(x=0,3,6,7) melt-spun alloys. For x=0, the alloy is entirely composed of grains of >100 nm, which are attributed to the α-Fe phase based on the XRD pattern. On the other hand, for x=7, large grains of >100 nm and also nanograins of <10 nm are observed: The large grains are considered to be responsible for the observed peak of α-Fe in the XRD pattern. For the alloys with x=3,6, nanograins of 10-30 nm are found to be distributed in the amorphous matrix. The alloy with x=6 contains more nanograins than that with x=3. The nanograins are considered to be the α-Fe phase. With increasing the B concentration from (b) to (d), the amount of nanocrystalline α-Fe is found to increase. Based on the above observations, nanocrystalline magnets with high (Fe,Co) concentration have been obtained by controlling the B/P concentration. It is understood that a high GFA is obtained by the simultaneous additions of B and P.

The results of the magnetic properties of $(Fe,Co)_{87}(P_{7-x}C_1B_x)_{13}$(x=0,3,6,7) are listed in table 1. When compared at the same (Fe, Co) concentration, the B addition is found to be effective in the improvement of the saturation magnetization. The nanocrystalline α-Fe grains dispersed in the amorphous matrix by the B addition (figure 2) is considered to be responsible for the high magnetizations: With increasing the B concentration from (b) to (d), the relative amount of nanocrystalline α-Fe is found to increase, which explains the increase of the magnetization. For $(Fe,Co)_{87}(P_1C_1B_6)_{13}$ and $(Fe,Co)_{87}(C_1B_7)_{13}$ alloys, the saturation magnetization reaches 1.85 T, which is comparable to commercial nanocrystalline soft materials (NANOPERM). On the other hand, the coercivity is also dependent on the B concentration. In general, the coercivity decreases with reducing the grain size d as d^{-6} [10], the lower coercivities obtained for the alloys with simultaneous B and P additions is attributed to the small grain sizes as seen in figure 2. The smallest coercivities are obtained for $(Fe,Co)_{87}(P_4C_1B_3)_{13}$ and $(Fe,Co)_{87}(P_1C_1B_6)_{13}$ when both B and P are added, and their values are reasonable for the observed grain sizes of ~ 30 nm.

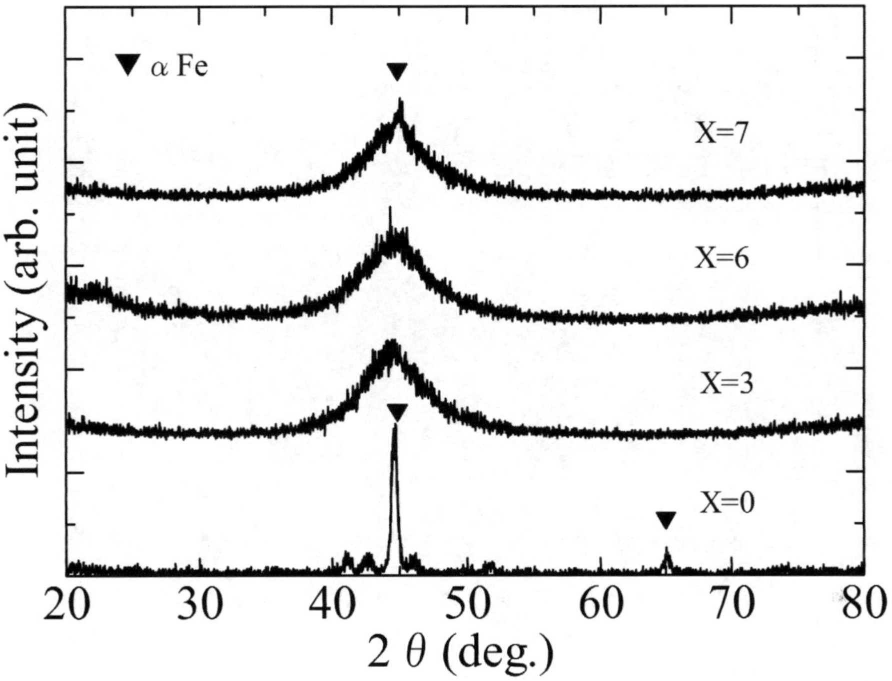

Figure 1. XRD patterns of $(Fe,Co)_{87}(P_{7-x}C_1B_x)_{13}$(x=0,3,6,7) melt-spun alloys

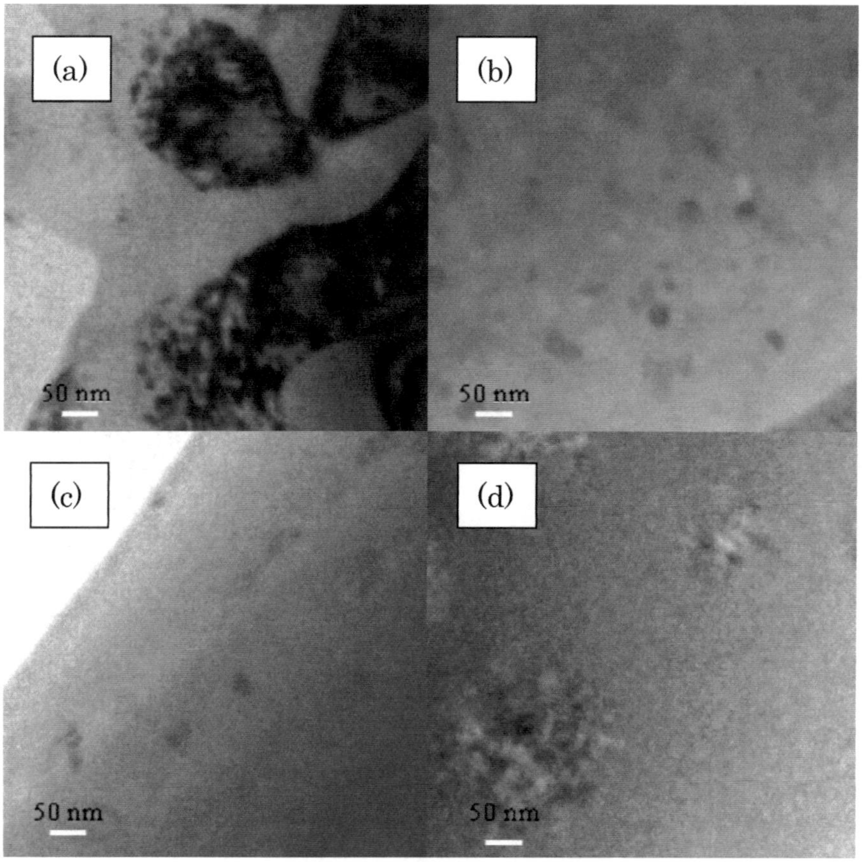

Figure 2. TEM images of $(Fe,Co)_{87}(P_{7-x}C_1B_x)_{13}$ (x=0,3,6,7) melt-spun alloys

(a) x = 0, (b) x = 3, (c) x = 6, (d) x = 7.

Table 1. Magnetic properties of $(Fe,Co)_{87}(P_{7-x}C_1B_x)_{13}$ (x=0,3,6,7) melt-spun alloys.

Composition(at%)	Bs(T)	Hc(A/m)	grain size(nm)
$(Fe, Co)_{87}(P_7C_1)_{13}$	1.70	>2400	>100
$(Fe, Co)_{87}(P_4C_1B_3)_{13}$	1.71	79	10～30
$(Fe, Co)_{87}(P_1C_1B_6)_{13}$	1.85	71	10～30
$(Fe, Co)_{87}(C_1B_7)_{13}$	1.86	97	>100 and <10

4. Conclusions

In the present work, high magnetization nanocrystalline soft materials of Bs~1.85 are obtained at the compositions of $(Fe,Co)_{87}(P_1C_1B_6)_{13}$ and $(Fe,Co)_{87}(C_1B_7)_{13}$. The B addition is found to increase the saturation magnetization through increasing the volume fraction of nanocrystalline α-Fe. The simultaneous additions of B and P are found to be effective in the grain refinement, resulting in low coercivity.

Acknowledgment

This work was supported by "Academic Frontier" Project for Private Universities: Matching Fund Subsidy from MEXT (Ministry of Education, Culture, Sports, Science and Technology), 2006-2010.

References

[1] Makino A, and Kubota K, *The 163th Topical Symposium of the Magnetic Society of Japan*

[2] Herzer G and Warlimont H 1992 *Nanostruct.Mater.* **1** 263

[3] McHenry M E, Willard M A, and Laughlin D E 1999 *Mater.Sci.* **44** 291

[4] Yoshizawa Y, Oguma S, and Yamauchi K 1988 *J. Appl. Phys.* **64** 6044

[5] Inoue A, Park R E 1996 *Mater. Trans.,JIM* **37** 1715

[6] Ohta M, and Yoshizawa Y 2008 *Hitachi metals technical review.* **24** 22

[7] Makino A, Hatanai T, Naitoh Y, Bitoh T, Inoue A, and Masumoto T 1997 *IEEE Trans. Magn.* **33** 3793

[8] Willard M A, Huang M Q, Laughlin D E, McHenry M E, Cross J O, Harris V G, and Franchetti C 1999 *J.Appl. Phys.* **85** 4421

[9] Inoue A, Takeuchi A, Zhang T. Murakami A, and Makino A 1996 *IEEE.Trans.Magn.* **32** 4866

[10] Herzer G 1997 *Handbook of Magnetic Materials* **10** 415

CH functionalization of heteroaromatic compounds by transition metal catalysis

Shunsuke Tanba, Taiki Fujiwara, Daiki Monguchi, Atsunori Mori

Department of Chemical Science and Engineering, Kobe University, 1-1 Rokkodai, Nada, Kobe 657-8501, Japan

amori@kobe-u.ac.jp

Abstract. Transition metal-catalyzed CH functioanlization of thiazoles and thiophenes are carried out. The reaction of thiophene with aryl halide in the presence of a palladium catalyst underwent the CC bond forming reaction at the CH bond of thiophene. By employing the reaction head-to-tail-type oligothiophene is synthesized in a stepwise manner. When several azoles are treated with secondary amines and amides in the presence of a copper catalyst, oxidative CH-NH coupling took place to form the carbon-nitrogen bond.

1. Introduction

CH functionalization by forming a carbon-carbon or carbon-heteroatom bond attracts considerable attention in organic synthesis recently.[1] The advantage of the CH functionalization is the simplicity of the total process. Although the related reaction such as the cross coupling of main group organometallics with organic halides furnishes the similar product,[2] the metallic species such as Grignard, boron, silicon, and tin have to be prepared with the corresponding compound bearing CX or CH moiety. We have been engaged in the development of coupling reactions at the CH bond of several heteroaromatic compounds such as thiazoles and thiophenes and shown that CH coupling of thiazole at the 2-position, which is the functionalization of acidic CH bond, takes place to afford the corresponding arylated product by Pd/Cu catalyst system.[3] (eq 1) It is also shown to undergo by the catalysis of palladium the coupling at the electron enriched carbon such as the 5-position of thiazole or the carbon atom of thiophenes adjacent to the sulfur atom.[4] (eq 2) In addition, oxidative homocoupling of thiophenes and thiazoles (5-position) at the CH bond is shown to occur in the presence of a palladium catalyst leading to the corresponding bithiazoles and bithiophenes.[5] Worthy of note in the CH cross coupling and homocoupling is that the tolerance of the carbon-bromine bond on the heteroaromatic ring to remain the halogen bond after the reaction and thereby allows the obtained coupling product toward further transformation reactions. Accordingly, when the reaction of an organohalogen compound bearing both carbon-bromine and carbon-iodine bonds is designed for CH functionalization, the reaction takes place only at the carbon-iodine bond, while the carbon-bromine bond remains. If it is possible to transform the bromo group into iodide, the product can be subjected to another CH functionalization reaction.

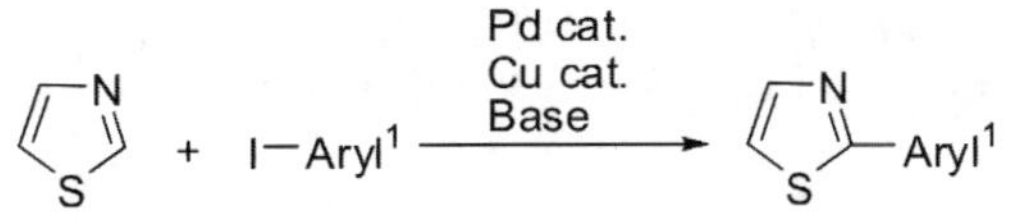

(1)

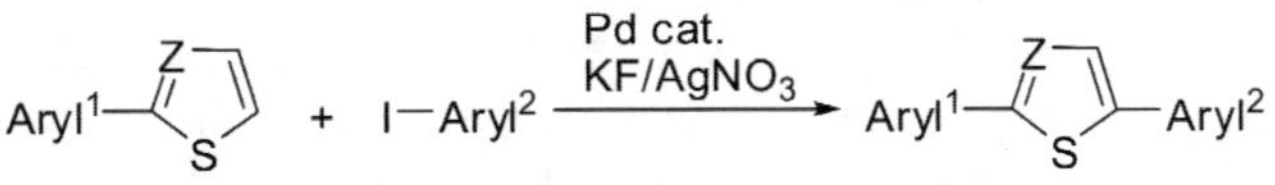

(2)

Z = N or CH

On the other hand, introduction of heteroatoms into heteroaromatic compound also attracts considerable attention since a wide range of heteroaromatic molecules bearing a heteroatom substituents are found biologically active compounds and advanced organic materials.[7] Transition metal catalysis is a powerful tool for the preparation of such compounds because substitution reaction at the sp^2 carbon-halogen bond with amines and other heteroatoms is established by Buchwald, Hartwig, and Koie.[8-10] By contrast, CH functionalization of heteroaromatic compounds with heteroatom is still an unexploited area and no example of CH-NH coupling of azoles has not been shown so far to the best of our knowledge.

Herein, we describe our recent progress on the CH functionalization reactions, which involves a new strategy to constitute iterative thiophene-thiophene unit with palladium-catalyzed CH coupling reaction and oxidative amination of azoles via CH-NH coupling catalyzed by a copper(II) salt.[11-12]

2. Results and discussion

2.1. Oligothiophene synthesis with iterative CH functionalization and halogen exchange

Since carbon-bromine bond is completely intact under the conditions of the palladium-catalyzed CH coupling reactions we have developed, the reaction of 2-bromothiophene at the CH bond with the carbon-iodine bond of another thiophene would induce CH-CI coupling, while carbon-bromine bond is intact. We envisaged the reaction of 2-bromo-3-hexylthiophene with a 5-substituted 3-hexyl-2-iodothiophene in the presence of $AgNO_3$/KF and palladium catalyst leading to the head-to-tail-type bithiophene. If it is possible to undergo halogen exchange of the produced bromide **3** into the corresponding iodide **4**, the product would be subjected to the further CH coupling reaction. Iterative CH coupling and halogen exchange sequence would lead to oligothiophenes in a stepwise manner, accordingly. The synthetic strategy is illustrated in Scheme 1.

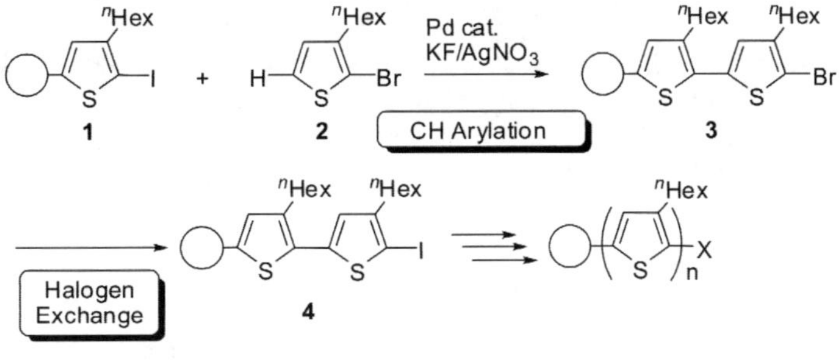

Scheme 1

We first studied the reaction conditions of halogen exchange with a copper salt. The halogen exchange reaction of aromatic bromide into iodide is shown to occur by Buchwald in the presence of a copper catalyst.[13] Thus, we applied the reaction conditions to the bromothiophene **2**. Although the reaction occurred in a reasonable yield after optimization of conditions, it was found that complete conversion of bromide into iodide was not achieved. Since the difficulty of the separation of iodide from the corresponding bromide by chromatography, the reaction would not be employed for the iterative oligothiophene synthesis. After screening the reaction conditions we found that the use of excess amount of CuI at the elevated temperature shown by Yamashita allowed to show the superior conversion.[14] A higher conversion (88%) was achieved when the reaction was carried out at 150 °C for 8 h under heterogeneous conditions employing 10 equiv of CuI. The reaction with a mixture of CuI (5 eq) and LiI (5 eq), which allows the reaction in DMSO in a homogeneous manner, afforded the corresponding iodide in 94% conversion. Almost complete conversion was achieved by employing CuI and LiI (10 equiv each). The results are summarized in Table 1.

Table 1. Cu(I) mediated halogen exchange reaction of 2-bromo-3-hexylthiophene (**2**)[a].

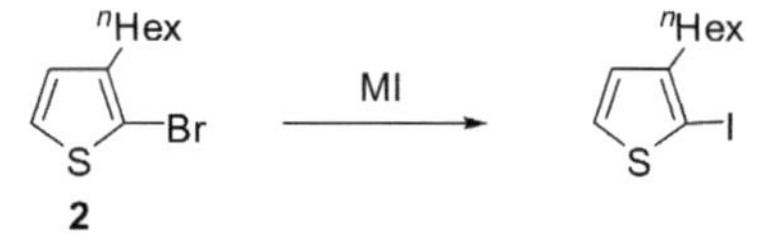

entry	CuI/ eq	MI (eq)	solvent (mL/mmol)	temp/ °C	time/ h	%conv.
1	0.1	NaI (3)	1,4-dioxane	110	24	78
2	5.0	-	DMSO	150	2	81
3	10.0	-	DMSO	150	8	88
4	1.0	LiI (4)	DMSO	100	4	14
5	5.0	LiI (5)	DMSO	100	4	9
6	5.0	LiI (5)	DMSO	150	25	94
7	10.0	LiI (10)	DMSO	150	25	97
8	2	LiI (8)	DMSO	150	13	64
9	1	LiI (9)	DMSO	150	13	33

[a] The reaction was carried out under a nitrogen atmosphere. Progress of the reaction was monitored by GC analysis of the reaction mixture.

With the optimum halogen exchange reaction condition in hand, stepwise reaction leading to a head-to-tail oligothiophene was performed with CH coupling/halogen exchange sequence. The reaction of **2** with ethyl 4-iodobenzoate (**6**) in the presence of a palladium catalyst and $AgNO_3$/KF afforded the corresponding coupling product **7** in 86% yield. The halogen exchange reaction of **7** under the optimized condition with LiI-CuI afforded the corresponding iodide **8** in 91% yield. Repeating further iterative CH arylation and halogen exchange sequence lead to the oligothiophene tetramer iodide **11** as shown in Scheme 2.

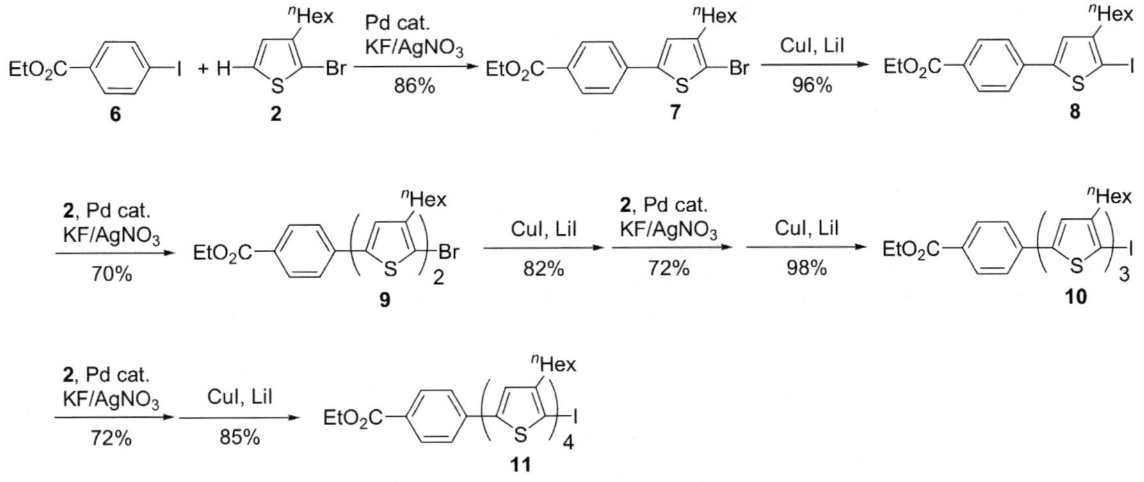

Scheme 2

2.2. CH-NH coupling of azoles

Benzothiazole (**12**) and *N*-methylaniline (**13**) was first chosen as a representative substrate and careful studies on the reaction conditions are carried out. When copper(II) acetate was employed as a catalyst species (20 mol%), the reaction was found to proceed under oxidative conditions in xylene at 140 °C in the presence of sodium acetate to afford the amination product **14** in 82% yield. Copper(II) was found to be a effective catalyst and the use of $CuCl_2$ also underwent the reaction with a slightly lower yield, while the reaction in the presence of 20 mol% of CuI resulted in no reaction. Other metal catalysts such as palladium(II) and iron(III) was also found to be ineffective. It is important to carry out the reaction under oxidative conditions. Indeed, the reaction under an oxygen atmosphere resulted in giving high yield, however, the yield was found to decrease (46%) under an aerobic condition and no reaction took place under a nitrogen atmosphere.

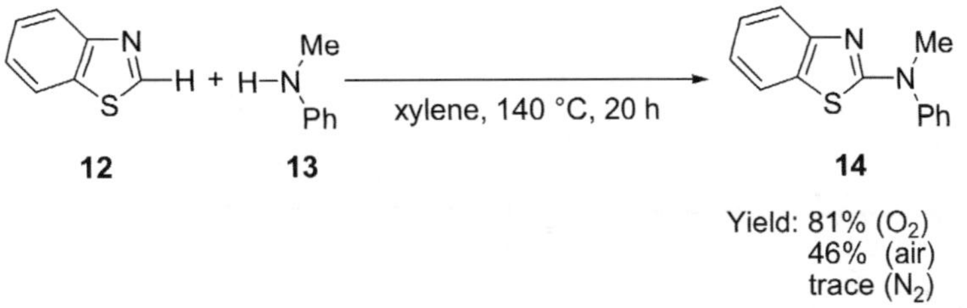

Table 2 summarizes the reaction of several azoles with amines. Benzothiazole reacted with diphenylamine and a sulfonamide, *N*-tosyl-*n*-propylamine in moderate to good yields. The reaction of 4,5-dimethylthiazole also proceeded, while 4-methylthiazole resulted in no reaction suggesting that an azole bearing the CH bond adjacent to the sulfur atom was not available. Several secondary aliphatic amines such as piperidine, and diethylamine reacted with benzoxazole in good to excellent yields. It was also found that the reaction of benzoimidazole with *N*-methylaniline occurred to afford the corresponding amination product in 51% yield.

Table 2. Copper-catalyzed CH-NH coupling of azoles with amines[a]

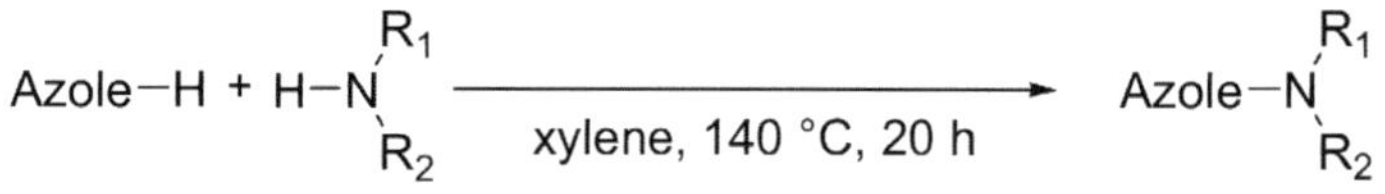

entry	Azole–H	amine	product	yield (%)[b]
1	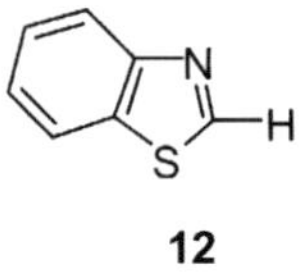 **12**	H-NMePh **13**	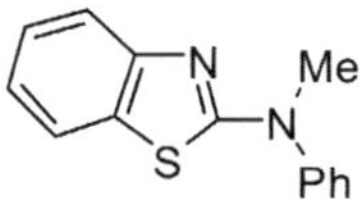 **14**	81
2	**12**	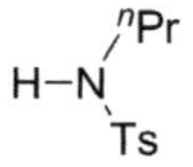	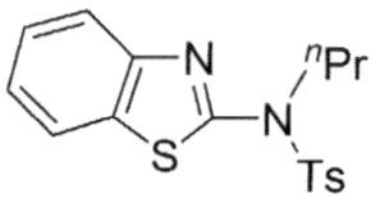	65
3	**12**	H-NPh_2	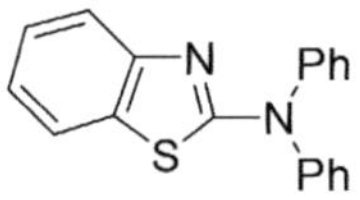	47
4	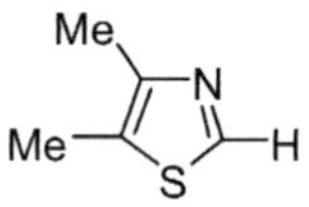	**13**	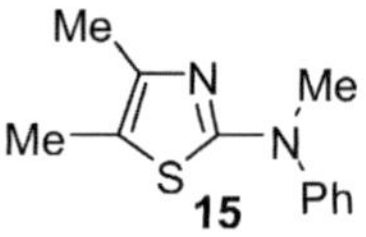 **15**	73
5	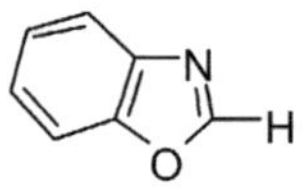	**13**	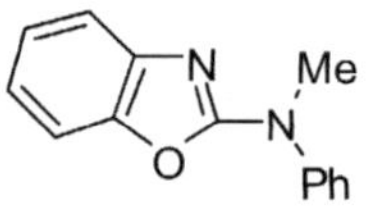	71
6	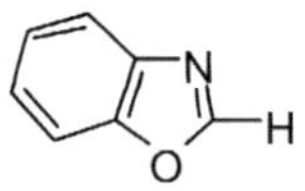	H-NPh_2	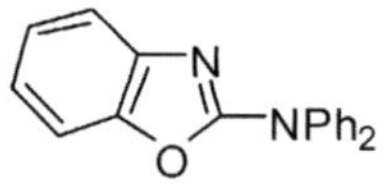 **16**	66
7	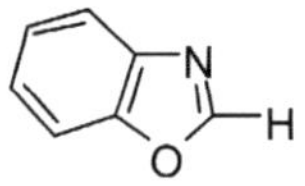	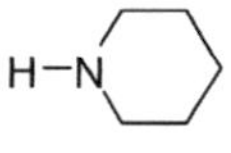	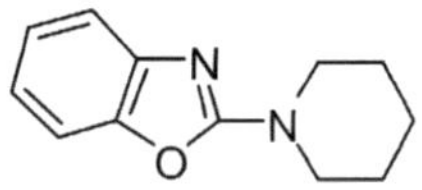	72
8	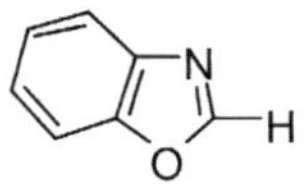	H-NEt_2	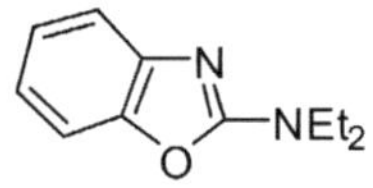	47
9	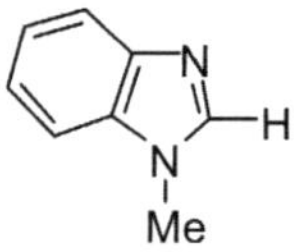	**13**	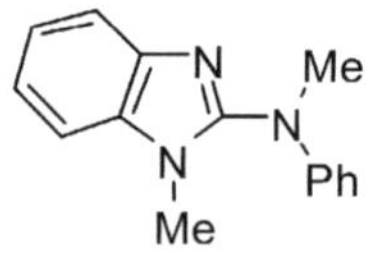	51

[a] The reaction was carried out with azole (1 eq), amine (4 eq), $Cu(OAc)_2$ (20 mol%), PPh_3 (40 mol%), and NaOAc (4 eq) in xylene at 140 °C for 20 h under O_2. [b] Isolated yield.

3. Conclusion

In summary, CH functionalization of five-membered heteroaromatic compounds such as thiophene and azoles was described. CC bond-formation of thiophene and thiazole was shown to take place at the electron-enriched carbon atom of heteroaromatic ring in the presence of a palladium catalyst. On the other hand, CH-NH coupling of azoles proceeded at the 2-position, which is the electron-deficient CH bond, with secondary amines and amides. These reactions involve potential utilities for the preparation of advanced electronic materials and biologically important organic compounds.

4. Experimental section

General procedure for the palladium-catalyzed CH arylation.

To a mixture of $PdCl_2(PPh_3)_2$ (8.8 mg, 0.01 mmol), potassium fluoride (18.2 mg, 0.31 mmol), and 4-iodobenzoic acid ethyl ester **6** (69 mg, 0.25 mmol) in 3 mL of DMSO was added 2-bromo-3-hexylthiophene **2** (0.07 mL, 0.3 mmol) and stirred at 100 °C for 5 h. $AgNO_3$ (10.6 mg, 0.31 mmol) was then added in five portions with an hour interval. After cooling to room temperature, the mixture was passed through a Celite pad, which was washed with chloroform repeatedly. The organic layer was concentrated under reduced pressure to leave a crude oil, which was purified by chromatography on silica gel to afford 85 mg of **7** as a yellow solid (86%).

General procedure for the halogen exchange reaction.

In a 100 mL of schlenk tube **7** (0.874 g, 2.21 mmol) was dissolved 25 mL of DMSO. To the solution were added copper iodide (I) (4.2 g, 22.1 mmol) and lithium iodide (2.96 g, 22.1 mmol) and stirred at 150 °C for 25 h. After cooling to room temperature, the mixture was quenched with water and it was passed through a Celite pad, which was washed with chloroform repeatedly. The organic layer was washed with $Na_2S_2O_3$ aq. and water three times, dried over anhydrous sodium sulfate, and concentrated under reduced pressure. The residue was purified by chromatography on silica gel to afford 0.941 g of **8** as an orange solid (96%).

Other reactions for the extension of the number of thiophene ring and CuI-LiI mediated halogen exchange were carried out in similar manners as shown above.

General procedure for the oxidative copper-catalyzed amination of heteroaromatic compounds.

A solution of $Cu(OAc)_2$ (7.3 mg, 0.04 mmol), PPh_3 (21 mg, 0.08 mmol), benzothiazole **12** (22 μL, 0.2 mmol), *N*-methylaniline **13** (88 μL, 0.8 mmol) and NaOAc (66 mg, 0.8 mmol) in 1 mL of xylene was stirred at 140 ºC for 20 h under O_2. After cooling to room temperature, the mixture was passed through a Celite pad, which was washed with chloroform repeatedly. The filtrate was washed with water three times. The organic layer was concentrated under reduced pressure to leave a crude oil, which was purified by column chromatography on silica gel to afford 39 mg of **14** as a colorless oil (81%).

***N*-(4,5-Dimethyl-thiazol-2yl)-*N*-methylaniline (15):** ^{1}H NMR (500 MHz, $CDCl_3$) δ 2.14 (d, J = 0.6 Hz, 3H), 2.18 (d, J = 0.6 Hz, 3H), 3.48 (s, 3H), 7.16-7.23 (m, 1H), 7.33-7.43 (m, 4H); ^{13}C NMR (125 MHz, $CDCl_3$) δ 11.1, 14.9, 40.0, 114.4, 124.7, 125.9, 129.7, 147.0, 166.4; IR (neat) 1131, 1287, 1374, 1410, 1516, 1597, 1678, 2851, 2917, 3035 cm^{-1}; HRMS found: m/z 218.0878. Calcd for 218.0878.

***N*-Benzoxazol-2-yl-diphenylamine (16):** ^{1}H NMR (500 MHz, $CDCl_3$) δ 7.03 (t, J = 7.6 Hz, 1H), 7.14 (t, J = 7.7 Hz, 1H), 7.18-7.22 (m, 3H), 7.27-7.37 (m, 8H), 7.41 (d, J = 7.6 Hz, 1H); ^{13}C NMR (125 MHz, $CDCl_3$) δ 109.4, 117.7, 122.2, 124.4, 126.2, 126.6, 129.6, 142.2, 142.8, 148.6, 160.3; IR (neat) 1006, 1209, 1242, 1366, 1456, 1495, 1557, 1625, 2852, 2922, 3057 cm^{-1}; HRMS found: m/z 286.1106. Calcd for 286.1103.

5. Acknowledgment

This work was partially supported by a Grant-in-Aid for Scientific Research on Priority Areas, "Advanced Molecular Transformation of Carbon Resources" and Special Coordination Funds for Promoting Science and Technology, Creation of Innovation Centers for Advanced Interdisciplinary Research Areas (Innovative Bioproduction, Kobe) both by Ministry of Education, Culture, Sports and Science and Technology (MEXT), Japan.

References

[1] Mori A and Sugie A 2008 *Bull. Chem. Soc. Jpn.* **81** 548.
[2] Diederich F and Stang P J, Eds 1998 *Metal-Catalyzed Cross-Coupling Reaction* Wiley VCH: Weinheim.
[3] Mori A, Sekiguchi A, Masui K, Shimada T, Horie M, Osakada K, Kawamoto M and Ikeda T 2003 *J. Am. Chem. Soc.* **125** 1700.
[4] Masui K, Mori A, Okano K, Takamura K, Kinoshita M and Ikeda T 2004 *Org. Lett.* **6** 2011.
[5] Masui K, Ikegami H and Mori A 2004 *J. Am. Chem. Soc.* **126** 5074.
[6] Takahashi M, Masui K, Sekiguchi H, Kobayashi N, Mori A, Funahashi M and Tamaoki N 2006 *J. Am. Chem. Soc.* **128** 10930.
[7] (a)Campeau L C, Stuart D R and Fagnou K 2007 *Aldrichimica Acta.* **40** 35. (b) Zificsak C A and Hlasta D J 2004 *Tetrahedron* **60** 8991. (c) Satoh T and Miura M 2007 *Chem. Lett.* **36** 200. (d) Seregin I V and Gevorgan V 2007 *Chem. Soc. Rev.* **36** 1173.
[8] Paul F, Patt J and Hartwig J F 1994 *J. Am. Chem. Soc.* **116**, 5969.
[9] Guram A S and Buchwald S L 1994 *J. Am. Chem. Soc.* **116**, 7901.
[10] Yamamoto T, Nishiyama M and Koie Y 1998 *Tetrahedron Lett.* **39** 2367.
[11] Masuda N, Tanba S, Sugie A, Monguchi D, Koumura N, Hara K and Mori A 2009 *Org. Lett.* **11** 2297.
[12] Monguchi D, Fujiwara T, Furukawa H and Mori A 2009 *Org. Lett.* **11** 1607.
[13] Klapars A and Buchwald S L 2002 *J. Am. Chem. Soc.* **124** 14844.
[14] Yamashita Y, Suzuki K and Tomura M 2003 *Synth. Met.* **133−134** 341.

Development of functional gold nanorods for bioimaging and photothermal therapy

T Niidome[1]
Faculty of Engineering, Kyushu University, Fukuoka 819-0395, Japan
Center for Future Chemistry, Kyushu University, Fukuoka 819-0395, Japan
PRESTO, Japan Science and Technology Agency, Kawaguchi 332-0012, Japan

E-mail: niidome.takuro.655@m.kyushu-u.ac.jp

Abstract. Gold nanorods have strong surface plasmon band at near-infrared light region, and are used as a photothermal converter. Since the near-infrared light penetrates into tissues deeply, it has been expected as a contrast agent for near infrared light bioimaging, a photosensitizer for photothermal therapy, and functional device for drug delivery system responding to near-infrared light irradiation. In this study, the surface plasmon bands of intravenously injected gold nanorods were monitored in the mouse abdomen using a spectrophotometer equipped with an integrating sphere, then we determined pharmacokinetics parameters of the gold nanorods after intravenous injection. Next, the PEG-modified gold nanorods were directly injected into subcutaneous tumors in mice, then, near-infrared pulsed laser light was irradiated the tumors. Significant tumor damage and suppression of the tumor growth was observed. We constructed targeted delivery system of the gold nanorods by modifying with a thermo-responsive polymer and a peptide responding to a protease activity. These modified gold nanorods are expected as functional nanodevices for photothermal therapy and drug delivery system.

1. Introduction

Gold nanorods, rod-shaped gold nanoparticles, have unique optical properties [1,2]. They show two surface plasmon bands corresponding to the transverse and longitudinal surface plasmon bands in the visible and the near infrared regions, respectively (Figure 1). The longitudinal band has a substantially larger extinction coefficient than the transverse band, and show efficient photothermal effect. The near infrared region (650 – 900 nm) is ideally suited for *in vivo* imaging and phototherapy due to minimum light absorption by intrinsic chromophores, hemoglobin, and water. Recently, we succeeded to prepare biocompatible gold nanorods by coating PEG chain [3,4]. Long lasting circulation of the PEG-modified gold nanorods was observed after intravenous injection. In this study, we tried to apply the gold nanorods as a contrast agent for near infrared light bioimaging, a photosensitizer for photothermal therapy, and a functional device for drug delivery system responding to near-infrared light irradiation and protease activity.

[1] To whom any correspondence should be addressed.

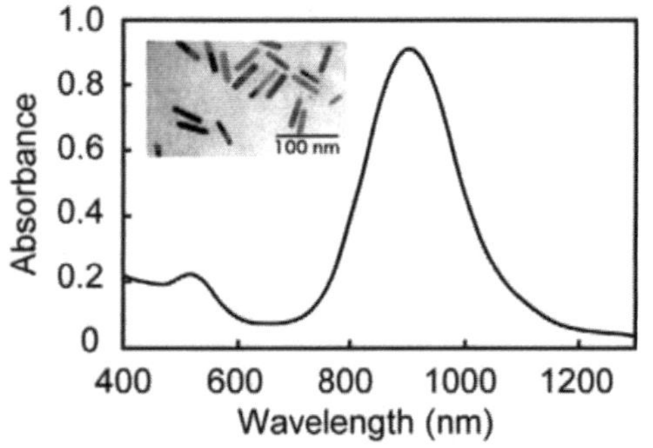

Figure 1. An absorbance spectrum of gold nanodods aqueous solution. *Inset*, a TEM image of the gold nanorods

2. Experimental

2.1. Gold nanorods

Gold nanorods, prepared with a modification of our previous methods [5], were supplied by Dai Nihon Toryo Co. Ltd. and Mitsubishi Materials Co. Ltd. Average length and width of as-prepared gold nanorods were 65 ± 5 nm and 11 ± 1 nm, respectively (aspect ratio: 5.9). PEG-modified gold nanorods were prepared as follows. A solution of gold nanorods containing CTAB was centrifuged at 14,000 *g* for 10 min, decanted, and resuspended in water to remove excess CTAB. 200 μl 5 mM $mPEG_{5,000}$-SH were added to 1 ml 1 mM gold nanorod solution. The mixture was stirred for 24 h at room temperature, and was dialyzed for 3 d.

2.2. Measurement of absorption spectra of gold nanorods

All animal experiments were performed according to Guidelines for Animal care and Use committee, Kyushu University. The anesthetized mouse was placed on a port of the integral sphere equipped with a conventional spectrophotometer (JASCO V-670), then, monochromatic light from the spectrophotometer was introduced into the abdomen of the mouse. The absorbance spectra of the mouse abdomen were obtained by scanning the wavelength of the light from spectrophotometer.

2.3. Preparation of NIPAM-modified gold nanorods

PEG-modified gold nanorods were coated with a silica layer using the modified Stöber method, which is based on the hydrolysis of tetraethyl orthosilicate (TEOS) in ethanol/water mixture in the presence of nanorods, which act as "seeds". Silane-coupling agent, 3-(methacryloxy) propyl triethoxysilane (MPS) was modified with the surface of the silica-coated nanorods. Next, silica-coated nanorods were modified with pNIPAM shells by precipitation polymerization. In the presence of MPS-modified silica nanorods, precipitation polymerization of NIPAM and *N,N'*-methylene bisacrylamide as a crosslinker was done. Finally, the silica templates were selectively etched using hydrofluoric acid.

2.4. Preparation of PEG-peptide-modified gold nanorods

Peptides were synthesized by Fmoc-chemistry manually. After deprotection and cleavage from resin, the peptides containing a cysteine residue were purified by RP-HPLC. *N*-hydroxy succinimide-terminated PEG (NOF Co., Ltd, MW ca. 5,000 Da) was reacted with amine of α-amino group of N-terminal amino acid in the peptide, then, resulting PEG-modified peptides were dialyzed for 3 d. Original gold nanorods solution containing CTAB was centrifuged at 14,000 *g* for 10 min, decanted, and re-suspended in water to remove excess CTAB. The PEG-modified peptides solution was added to gold nanorods suspension at various PEG:Au molar ratios. The mixtures were stirred for over night at room temperature and centrifuged at 14,000 *g* for 10 min, decanted, and resuspended in water to remove remaining CTAB and excess PEG-modified peptides. Zeta potential of PEG-peptide-modified gold nanorods were around neutral.

2.5. Photothermal tumor damage

For evaluation of tumor growth suppression, B-16 melanoma cells were injected subcutaneously into abdomen of mouse (BALB/c, male, 3 weeks, 10 - 13 g) and allowed to grow for 5 days, where the tumors reached a size of ~ 5 mm in diameter. The gold nanorods (50 μl in 5% glucose, 2 mM) were injected directly into the tumor, and then the tumor was irradiated with pulsed laser light (Nd:YAG 1,064 nm, 750 mW, 20 Hz, 3 min, ~ 5 mm beam diameter) within 10 min after the injection. In the case of intravenous injection, the gold nanorods (300 μl in 5% glucose, 2 mM) were injected. At 24 h after the injection, the tumor was irradiated with laser light.

3. Results and Discussion

3.1. Measurement of absorption spectra of gold nanorods

As shown in Figure 2, spectra of the gold nanorods could be observed at abdomen immediately after injection (*spectra a*) and 30 min after injection (*spectra b*). Baseline data were recorded just before the injection. Following the injection of PEG-modified gold nanorods (Figure 2A), characteristic peaks at around 900 nm were observed. Furthermore, because aggregation of the gold nanorods dramatically changes the longitudinal surface plasmon bands in the near infrared region, the peak intensities could be assigned to the amount of the gold nanorods dispersing in the abdomen without forming aggregates. Thus, it indicates that the PEG-modified gold nanorods circulated at least 30 min without forming aggregates. In the case of injection of phosphatidylcholine (PC)-modified gold nanorods [6], a surface plasmon band from the gold nanorods was observed at immediately after the injection (Figure 2B). However, at 30 min after injection of the PC-modified gold nanorods, no absorption peak was observed in the spectrum, indicating that circulation of gold nanorods in mice was not as good as that of the PEG-modified gold nanorods.

Next, the absorbance at 900 nm was continuously monitored after the intravenous injection. The injection of the PEG-modified gold nanorods resulted in an increase of absorption in the abdomen, which then reached a plateau. On the other hand, the injections of the PC-modified nanorods resulted in an increase of absorption, as in the case of PEG-modified gold nanorods; however, the absorption intensities immediately decreased. By single-exponential fitting of the decrease, half-lives of the PEG- and PC-modified gold nanorods were estimated to be 231 and 1.3 min, respectively. The PC-modified gold nanorods have slight positive surface charge. Positively charged nanoparticles tends to be recognized by reticuloendothelial system in liver and spleen, therefore, their half-life was shorter than the PEG-modified gold nanorods that have neutral surface. Such a surface chemical structure of the gold nanorods was important factor for stability in the circulation and biodistribution.

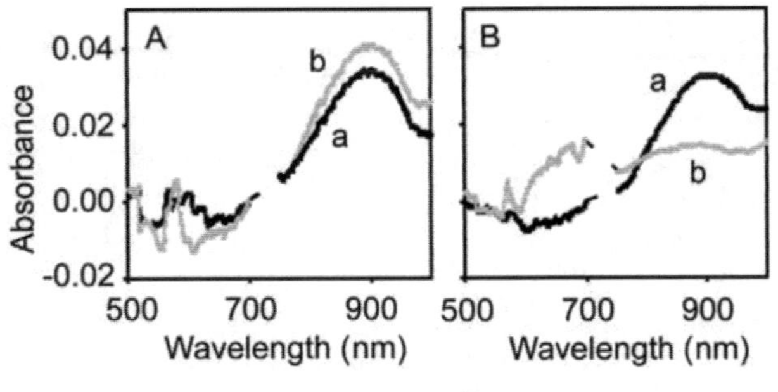

Figure 2. Absorption spectra from the abdomen of mice after intravenous injection of gold nanorods (300 μl of 2 mM Au atoms). Spectra immediately after injection (a) and 30 min after injection (b) of PEG- (A) and PC-modified (B) gold nanorods.

3.2. Photothermal tumor damage

We tried to perform photothermal cancer therapy of tumor bearing mice. To confirm the therapeutic effect of the photothermal system, we measured the tumor growth after the treatment (Fig. 3). Only in the case where both direct injection of gold nanorods and the irradiation of laser light resulted in a significant suppression of the tumor (Figure 3, *closed circles*), while the tumor growth in the control mice, that is, only those mice injected with gold nanorods or those mice treated with laser irradiation (Fig. 3, *open squares* and *closed squares*, respectively) showed no difference from the case of non-treated mice (Fig. 3, *closed triangles*) [7]. In the case of intravenous injection of gold nanorods, the suppression of the tumor growth was lower than the case of the direct injection (data not shown), indicating that the targeted delivery of the gold nanorods to the tumor tissue is an important key to improve the therapeutic effect.

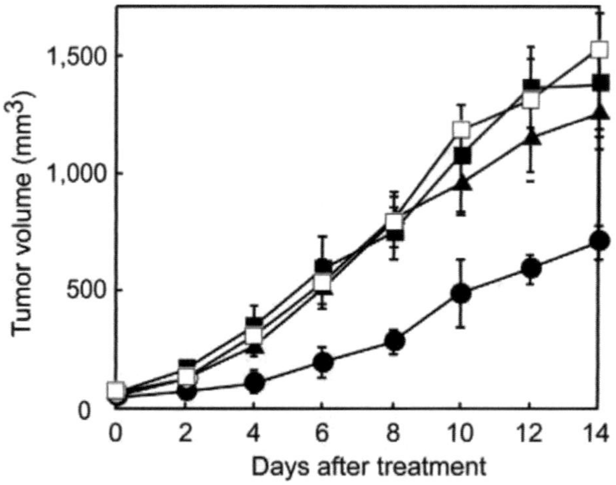

Figure 3. Tumor growth after photothermal treatment. *Closed circles* indicate the case of both direct injection of the gold nanorods and laser irradiation. *Open squares* indicate the case of direct injection of gold nanorods alone. *Closed squares* indicate the case of direct injection of 5% glucose followed by laser irradiation. *Closed triangles* indicate the case of non-treated mice. Data represent mean values for n= 6 and bars are standard deviations for the means.

3.3. Construction of targeting system of gold nanorods

To achieve targeted delivery of the nanorods to specific site, firstly we focused on thermoresponsive polymer (pNIPAM), that is, we expected that generated heat by photothermal effect of the gold nanorods induces phase transition of the pNIPAM layer to hydrophobic (Figure 4). The hydrophobic surface could interact with cellular membrane or extracellular matrix, therefore, the nanorods will accumulate to the irradiated site. We prepared hybrid nanogels (a single gold nanorod encapsulated in a PNIPAM nanogel) by colloid-template polymerization and silica etching [8]. pNIPAM-coated gold nanorods shrank considerably and rapidly when near infrared laser was irradiated (~4 W/cm^2). A cycle of shrinkage and swelling could be induced by switching the laser on and off. Next, we tried a feasible study on *in vivo* application. The gel modified gold nanorods were intravenously injected into mice, then, the right kidney was immediately irradiated using the NIR laser. Organs were subsequently collected, and the amounts of the gold nanorods in organs were evaluated by ICP-MS. When a solution of pNIPAM-coated gold nanorods was injected, the golds were distributed primarily in the blood (47%) and lungs (46%) after 10 min. The lower critical solution temperature of the pNIPAM layer was 34°C lower than body temperature, but they circulated in the blood within 10 min because of a slow phase transition response to a jump in environmental temperature. After irradiation of the right kidney for 10 min immediately after injection of nanogels, significant accumulation of the gold in the irradiated right kidney was observed. In that time, no gold was observed in the left kidney and the blood. pNIPAM-coated gold nanorods, which circulated in the blood after injection, could accumulate at the irradiated site due to induced rapid photothermal phase transition.

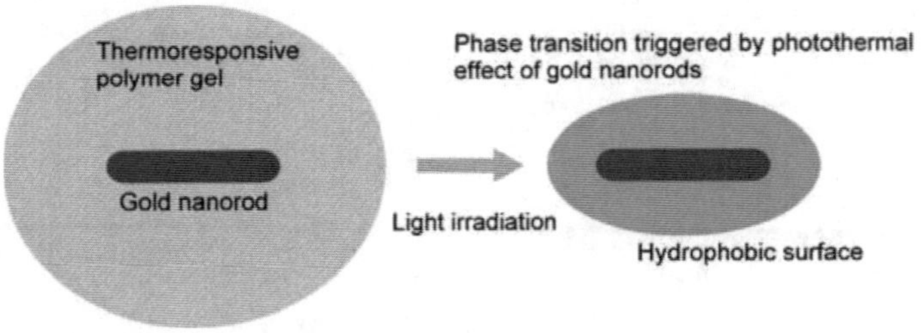

Figure 4. Shrinkage of pNIPAM gel layer on the gold nanorods due to its phase transition induced by photothermal effect of the gold nanorods.

Second, we tried another strategy that focused on protease specifically expressing in tumor tissue, that is, a substrate peptide (LGGSGRSANAILE-Cys) for urokinase-type plasminogen activator (uPA) was conjugated with PEG chain, then, the PEG-modified peptide was bound on the surface of the gold nanorods (Figure 5). When purified uPA was added to the PEG-peptide-modified gold nanorods, aggregation of the nanorods was confirmed by the absorption of the near infrared light. Decrease of the absorption depended on the amount of uPA and density of PEG-peptide on the nanorods. Since uPA is specifically expressed in tumor tissue, the PEG-peptide-gold nanorods would accumulate in the tumor due to the aggregation triggered by the site-specific enzyme activity.

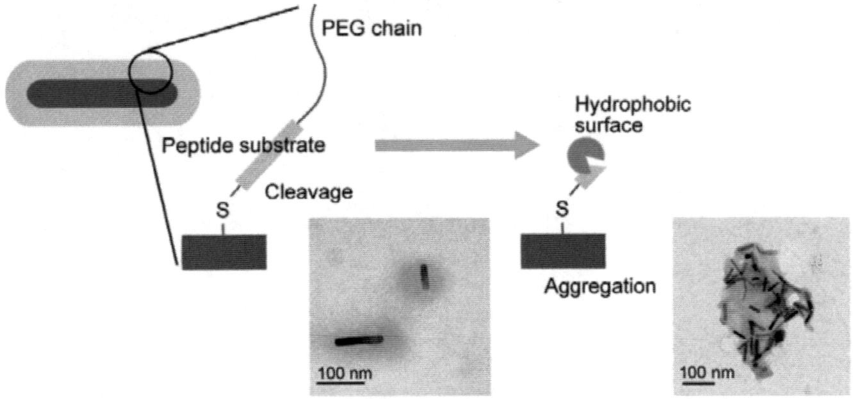

Figure 5. PEG-peptide modification of the gold nanorods and their aggregation triggered by enzymatic cleavage of the peptide chain.

4. Conclusion

We succeeded in direct monitoring of gold nanorods *in vivo* with an integrating sphere, then, half-lives of the gold nanorods in the circulation could be calculated. Second, we demonstrated photothermal cancer therapy by combining the injection of PEG-modified gold nanorods and near infrared laser irradiation. Finally, we constructed delivery systems responding to near infrared laser irradiation and protease activity. These functional systems will allow us to diagnose the tumors using the near infrared light and improve the therapeutic effect after intravenous injection of the gold nanorods.

Acknowledgments

This research was supported by a Grant-in-Aid for Scientific Research (B) (No. 19300172 and 18310069) from the Japan Society for the Promotion of Science (JSPS) and the Precursory Research for Embryonic Science and Technology (PRESTO) from the Japan Science and Technology Agency.

References

[1] Yu Y –Y, Chang S –S, Lee C –L and Wang C R C 1997 *J. Phys. Chem. B* **101** 6661

[2] Link S, Mohamed M B and El-Sayed M A 1999 *J. Phys. Chem. B* **103** 3073

[3] Niidome T, Yamagata M, Okamoto Y, Akiyama Y, Takahashi H, Kawano T, Katayama Y and Niidome Y 2006 *J. Control. Release* **114** 343

[4] Shiotani A, Mori T, Niidome T, Niidome Y and Katayama Y 2007 *Langmuir* **23** 4012

[5] Niidome Y, Nishioka K, Kawasaki H and Yamada S 2003 *Chem. Commun.* **18** 2376

[6] Takahashi H, Niidome Y, Niidome T, Kaneko K, Kawasaki W and Yamada S 2006 *Langmuir* **22** 2

[7] Niidome T, Akiyama Y, Yamagata M, Kawano T, Mori T, Niidome Y and Katayama Y 2009 *J. Biomater. Sci.-Polym. Ed.* **20** 1203

[8] Kawano T, Niidome Y, Mori T, Katayama Y and Niidome T 2009 *Bioconjugate Chem.* **20** 209

Nickel-catalyzed Reactions of Enone with Ethylene

A Nishimura [1] **T Haba** [1] **M Ohashi** [1,2] **and S Ogoshi** [1,3]

[1] Department of Applied Chemistry, Faculty of Engineering, Osaka University, Suita, Osaka 565-0871, Japan
[2] Center for Atomic and Molecular Technologies, Osaka University, Suita, Osaka 565-0871, Japan

ogoshi@chem.eng.osaka-u.ac.jp

Abstract. The reaction of (*E*)-1-phenylbut-2-en-1-one with ethylene in the presence of a catalytic amount of $Ni(cod)_2$ and PCy_3 at room temperature gave two kinds of three-component addition products; one is 1,6-enone composed of an enone and two ethylene molecules, and the other is 1,5-diketone composed of two enones and an ethylene. The reactions might proceed via oxidative cyclization of an enone and an ethylene with nickel(0).

1. Introduction

Formation of a carbon-carbon bond by the addition of C-H bonds across carbon-carbon unsaturated compounds, such as alkynes and alkenes, constitutes a particularly variable strategy, because it is one of the most atom-economical reactions to construct a carbon-carbon bond. For one thing, direct conjugate addition reactions of enolizable substrates, such as 1,3-dicarbonyl compounds to α,β-unsaturated carbonyl compounds, are employed for forming a carbon-carbon bond [1]. For another, transition metal catalyzed direct conjugate additions of terminal alkynes to enones have been reported as well (Scheme 1) [2-6].

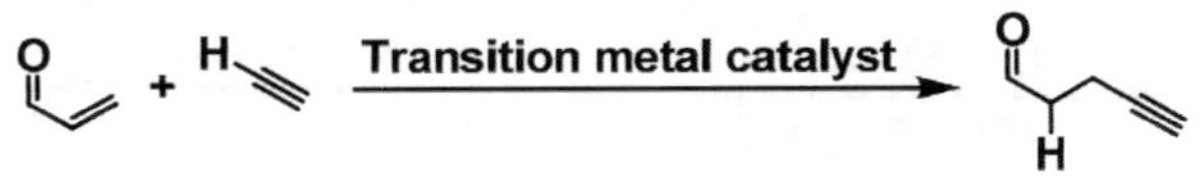

Scheme 1. Transition metal catalyzed direct conjugate addition reactions of terminal alkynes to enones.

In addition, construction of carbon-carbon bonds between enones and alkynes can also be achieved by nickel-catalyzed three-component cycloaddition reactions (Scheme 2). Ikeda *et al.* developed nickel-

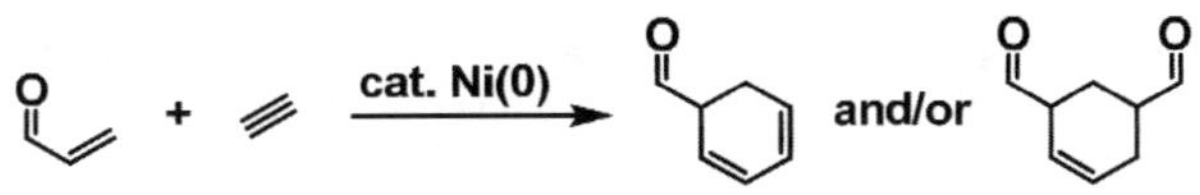

Scheme 2. Ni-catalyzed three-component cycloaddition reactions of enones with alkynes.

[3] To whom any correspondence should be addressed.

catalyzed reactions of an enone with two alkynes to afford cyclohexadiene derivatives [7; 8]. Nickel-catalyzed intramolecular reactions of an alkyne and two enones, yielding cyclohexene derivatives, were reported by Montgomery *et al.* [9].

In contrast, it has been believed to be difficult that an alkenyl group is introduced into the β-position of an enone by the addition of C-H bonds of simple alkenes as reported for terminal alkynes, since a C-H bond of simple alkenes is much less reactive than that of the terminal alkynes. We recently demonstrated the first example of a nickel-catalyzed direct conjugate addition reactions of simple alkenes, such as styrene, to enones, which is a very straightforward process to introduce an alkenyl group to the β-carbon of enones by the addition of a carbon–hydrogen bond (Scheme 3) [10; 11]. In this reaction, oxidative cyclization of alkenes and enones with nickel(0), giving a nickelacycle intermediate [12-16], was proposed as a key reaction step.

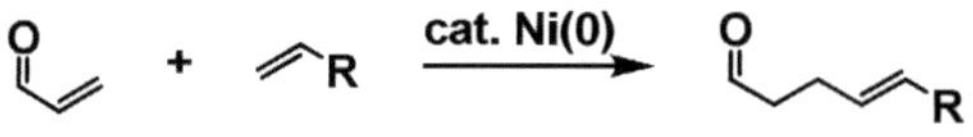

Scheme 3. Ni-catalyzed direct conjugate addition reactions of alkenes to enones.

As a part of investigation on the reactivity of enones with alkenes, we focused attention on a Ni-catalyzed reaction of enones with ethylene, since ethylene is one of the important industrial feedstock. To date, a variety of reactions have been developed to convert ethylene into more valuable chemicals, and the hydrovinylation reaction of alkenes mediated by transition metal hydride complexes is regarded as one of the most useful reactions for the introduction of ethylene into alkenes. However, the isomerization of terminal alkenes to internal alkenes also took place under the reaction conditions using transition metal hydride species (Scheme 4), therefore, a limited number of alkenes, such as vinylarenes, cyclic dienes, and norbornenes, are available for hydrovinylation reactions [17].

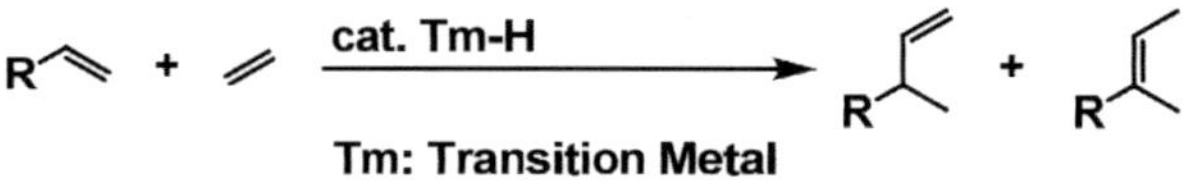

Scheme 4. Hydrovinylation reactions catalyzed by transition metal hydride species.

We, thus, investigated the reactivity of (*E*)-1-phenylbut-2-en-1-one (**1**) with ethylene in the presence of Ni(0) catalyst, resulting in the formation of three-component addition products [18]. To the best of our knowledge, this is the first example of the three-component addition reactions of enones with alkenes, which afford terminal alkenes [11; 19; 20].

2. Results and Discussion

In the presence of $Ni(cod)_2$ and PCy_3 (cod = 1,5-cyclooctadiene, Cy = c-C_6H_{11}), the reaction of **1** with 5 atm of ethylene at room temperature in C_6D_6 resulted in the formation of a mixture of 1,6-enone **2** and 1,5-diketone **4** in 7% and 41% yield, respectively (Table 1, Run 1) [21]. Both products **2** and **4** were three-component addition compounds; the enone **2** was composed of an enone and two ethylenes, whereas the ketone **4** consisted of two enones and an ethylene.

With elevating the reaction temperature to 80 °C, the yields of **2** and its isomer **3** increased and the formation of **4** was suppressed in 19% yield (Run 2). The yield of **2** was somewhat improved to 30% under a higher (8 atm) ethylene pressure conditions (Run 3). The reaction was carried out under a much higher ethylene pressure (30 atm) with much diluted solution of **1** (0.04 M). The selectivity of **2** and **3** toward **4** was, as a result, found to be improved, and finally, a mixture of **2**, **3**, and **4** was isolated in 74% yield (Run 4).

Table 1. Reaction of **1** with ethylene in the presence of Ni(cod)$_2$ and PCy$_3$.

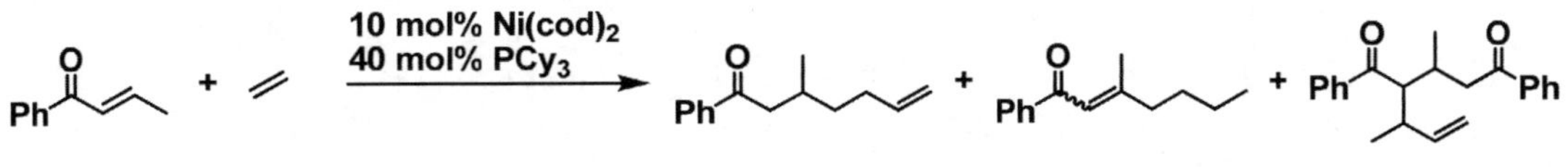

1 2 3 4

Run	Ethylene	Temp/°C	Time/h	Yield/% 2	3	4
1[a]	5 atm	25	72	7	-	41
2[a]	5 atm	80	17	20	3	19
3[a]	8 atm	80	17	30	5	18
4[b]	30 atm	100	15	41	18	13

[a][1] = 0.4 M, in C_6D_6, NMR yield.
[b][1] = 0.04 M, in toluene, isolated yield.

A plausible reaction mechanism, although any nickelacycle intermediates were not directly observed at all, is depicted in Scheme 5. The oxidative cyclization of an enone and ethylene with a nickel(0) species generates a η^3-oxaallylnickelacycle intermediate (**A**). The formation of the corresponding nickelacycle by the intramolecular oxidative cyclization of an enone and an alkene with nickel(0) has been reported [13]. The direct conjugate addition product (**B**), the analogue of which was obtained by employing styrene instead of ethylene as an olefin partner [10], was not generated at all, indicating that β-hydrogen elimination from **A** did not take place.

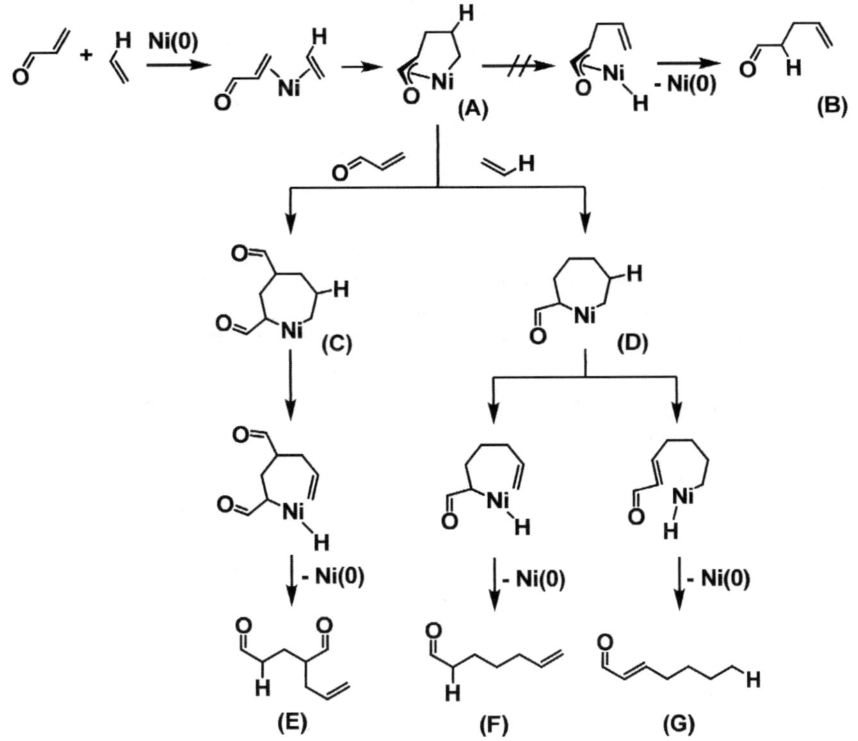

Scheme 5. A plausible mechanism for the formation of three-component coupling products.

Instead, insertion of either an enone or an ethylene to **A** generates a seven-membered nickelacycle intermediate (**C**) or (**D**), respectively. The seven-membered nickelacycle intermediate **C** would undergo β-hydrogen elimination followed by reductive elimination to give a 1,5-diketone product (**E**). It should be mentioned that the seven-membered nickelacycle intermediate **C** might be formed by the conjugate addition of **A** to an enone, since the conjugate addition of a six-membered nickel enolate to an enone proceeded [22; 23].

Likewise, the other seven-membered intermediate **D** would also undergo β-hydrogen elimination followed by reductive elimination, giving a mixture of 1,6-enone compound (**F**) and its isomeric 1,2-enone (**G**). As an alternative reaction path to generate **D**, the oxidative cyclization of $(\eta^2\text{-}CH_2{=}CH_2)_2Ni(PCy_3)$ followed by the insertion of an enone into nickel–carbon bond is also possible. In fact, at the end of the reaction, the formation of $(\eta^2\text{-}CH_2{=}CH_2)_2Ni(PCy_3)$ was confirmed [24].

3. Conclusions

We demonstrated the first example of the reaction of an enone with two ethylene molecules in the presence of $Ni(cod)_2$ and PCy_3. This reaction product, a 1,6-enone with a terminal alkenyl moiety, can be regarded as a cross-trimerization compound composed of alkenes with ethylene. In addition, at a lower temperature, the reaction of two enones with an ethylene gave a 1,5-diketone compound as a major product.

References

[1] Perlmutter P 2002 *Tetrahedron Organic Chemistry Series 7: Conjugate Addition Reactions in Organic Chemistry* 63

[2] Nikishin, G I and Kovalev, I P 1990 *Tetrahedron Lett.* **31** 7063

[3] Picquet M, Bruneau C, and Dixneuf P H 1999 *Tetrahedron* **55** 3937

[4] Chang S, Na Y, Choi E, and Kim S 2001 *Org. Lett.* **3** 2089

[5] Lerum R V and Chisholm J D 2004 *Tetrahedron Lett.* **45** 6591

[6] Chen L and Li C –J 2004 *Chem Commun.* 2362

[7] Ikeda S, Mori N and Sato Y 1997 *J. Am. Chem. Soc.* **119** 4779

[8] Mori N, Ikeda S and Sato Y 1999 *J. Am. Chem. Soc.* **121** 2722

[9] Seo J, Chui H M P, Heeg M J and Montgomery J 1999 *J. Am. Chem. Soc.* **121** 476

[10] Ogoshi S, Haba T and Ohashi M 2009 *J. Am. Chem. Soc.* **131** 10350

[11] A nickel-catalyzed reaction of enones with ethylene in the presence of a trialkylsilyl triflate and triethylamine to give corresponding silyl ethers incorporating one ethylene molecule: Ho C-Y, Ohmiya H and Jamison T F 2008 *Angew. Chem., Int. Ed.* **47** 1893

[12] Ogoshi S, Oka M and Kurosawa H 2004 *J. Am. Chem. Soc.* **126** 11802

[13] Ogoshi S, Ueta M, Arai T and Kurosawa H 2005 *J. Am. Chem. Soc.* **127** 12810

[14] Ogoshi S, Tonomori K-I, M Oka and Kurosawa H 2006 *J. Am. Chem. Soc.* **128** 7077

[15] Ogoshi S, Ikeda H amd Kurosawa H 2007 *Angew.Chem., Int. Ed.* **46** 4930

[16] Ogoshi S, Arai T, Ohashi M and Kurosawa H 2008 *Chem. Commun.* 1347

[17] RajanBabu T V 2003 *Chem. Rev.* **103** 2845 and references cited therein

[18] Ogoshi S, Nishimura A, Haba T and Ohashi M 2009 *Chem. Lett.* **38** 1166

[19] [2 + 2 + 2] Cycloadditions of electron-deficient olefins with norbornadienes have been reported. Lautens M, Edwards L G, Tam W and Lough A J, 1995 *J. Am. Chem. Soc.* **117** 10276

[20] In a similar reaction with alkenes, Hoberg *et al.* have reported a nickel-catalyzed three-component addition reaction of a phenyl isocyanate with two ethylenes. Hoberg H and Hernandez E 1986 *J. Chem. Som., Chem. Commun.* 544

[21] An unidentified isomer of **4** (less than 1%) was also detected in ^1H and ^{13}C NMR spectra.

[22] Ogoshi S, Nagata M and Kurosawa H 2006 *J. Am. Chem. Soc.* **128** 5350

[23] Tamaki T, Nagata M, Ohashi M and Ogoshi S 2009 *Chem. -Eur. J.* **15** 10083

[24] Pörschke K-R 1989 *J. Am. Chem. Soc.* **111** 5691

Magnetoresistance oscillation at charge neutrality point in monolayer graphene due to potential fluctuation

Ryuta Yagi[1,3], Seiya Fukada[1], Hiroaki Kobara[1], Yumi Shintani[1], Norio Ogita[1,3] and Masayuki Udagawa[1,3]

1 Graduate School of Advanced Sciences of Matter (AdSM), Hiroshima University, Kagamiyama 1-3-1, Higashi-Hiroshima, 739-8530, Japan

2 Graduate School of Integrated Art and Science, Hiroshima University, Kagamiyama 1-7-1, Higashi-Hiroshima, 739-8521, Japan

3 Institute for Advanced Material Research, Kagamiyama 1-3-1, Higashi-Hiroshima, 739-8530, Japan

E-mail: yagi@hiroshima-u.ac.jp

Abstract.
We have studied magnetoresistance fluctuation of monolayer graphene at the minimum conductivity point. We found that peaks of the fluctuation could be assigned to a few series of oscillation periodic in $1/B$. Temperature dependence of the amplitude indicated that the oscillation is due to Shubnikov-de Haas effect arising from potential fluctuation. From the oscillation, we have estimated typical magnitude of the potential fluctuation and size of the local domain.

1. Introduction

Since realization of graphene FET [1], a number of theoretical and experimental studies were carried out to date. An electron and a hole band of graphene have linear dispersion relation near K and K' point. The electronic states are described by an equation of massless Dirac fermions. One of the characteristic features stemming from the massless Dirac fermions is Landau quantization. The energy of Landau levels has square dependence on magnetic field and Landau level index N. Moreover, zero-mode, a Landau level with $N = 0$, is formed at the Dirac point. It is not formed in normal metals and semiconductors. The particular Landau quantization results in the half-integer quantum Hall effect: Hall conductivity is quantized to $\sigma_{xy} = 4(N_{\rm qhe} + 1/2)e^2/h$ [2][3]where $N_{\rm qhe}$ is integer (0, ±1, ±2, ±3,$\cdots$). This essentially stems from the fact that massless Dirac fermions acquire geometrical phase of -π for a cyclotron orbit. Magnetoresistance near a charge neutrality point is qualitatively different from those with larger carrier densities. Positive magnetoresistance appearing at low magnetic fields is followed by saturation accompanied by a small fluctuation. In much larger magnetic field, magnetoresistance is positive[4]. The above mentioned behavior near the minimum conductivity point is distinct from the behavior observed at larger carrier densities where only Shubnikov-de Haas oscillation is discernible. In this paper we report experimental study about the fluctuating magnetoresistance at the minimum conductivity point, whose origin is currently unknown.

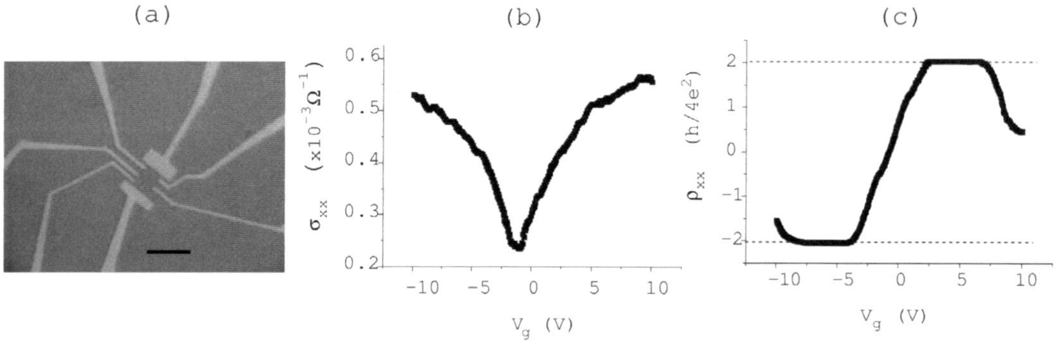

Figure 1. (a) Photo micrograph of a monolayer graphene device. The bar in the figure indicates 10 μm. (b) Gate voltage (V_g) dependence of conductivity σ_{xx} measured at $B = 0$ T and $T = 4.2$ K. (c) V_g dependence of ρ_{xy}. A plateau for half-integer quantum Hall effect with $N_{\rm qhe} = \pm 1$, *i.e.*, filling factor, $\nu = \pm 4$, can be seen.

2. Experimental

Our graphene flakes were prepared by micro-mechanical exfoliation [1] of Kish graphite on a heavily doped p-type Si substrate covered with 3000 Å of SiO_2. We selected monolayer graphene flake using optical micrograph, identifying number of layers by analyzing green channel of RGB image of the graphene flake using a relation between color intensity and number of layer, which was calibrated by Raman spectroscopy in advance. Electric leads for transport measurements were connected using electron beam lithography and vacuum evaporation. Figure 1(a) shows a photo micrograph of our graphene device. Contacts were formed by depositing of 3 nm of chromium and 17 nm of gold successively. The devices were cooled down to $T = 4.2$ K. Magnetic field was applied to the device with superconducting solenoid up to $B = 7.6$ T. Resistivity measurement was carried out by a standard lock-in technique. Figure 1(b) shows gate voltage dependence of conductance at zero-magnetic field. The conductivity took the minimum at about $V_g =$ -1.2 V. Using equations in Ref. [5], we estimated mobility to be 4.2×10^3 cm$^2/Vs$, residual charge density to be 3.5×10^{10} cm^{-2}, and impurity density $n_{\rm imp}$ to be 1.0×10^{11} cm^{-2} at the vicinity of charge neutrality point. Our graphene was confirmed to be monolayer by the appearance of quantum Hall plateau in Vg dependence of ρ_{xy} at B=7.6 T (Fig. 1(c)), which indicated half-integer quantum Hall effect with $N_{\rm qhe} = \pm 1$.

3. Result and Discussion

Magnetoresistance at the minimum conductivity point is shown in Figure 2, showing characteristic structure: the resistance increased rapidly from $B = 0$ to 2 T, which is followed by a saturation accompanied by slight negative magnetoresistance [6][7]. In addition to the background structure, fluctuating magnetoresistance can be seen. The fluctuation was reproducible from sweep to sweep, however, they were highly sample-dependent as shown in the inset. This oscillation is observed at the gate voltages only near the charge neutrality point. These two facts, sample- and V_g- dependence, indicated a possible relevance of the origin to the inhomogeneity induced potential fluctuation.

In order to study the nature of the fluctuation, we analyzed the periodicity. We could assign the peaks to three major series of oscillations A, B and C, periodic in 1/B as shown by the arrows in Fig. 2. The period of the oscillation was $\Delta(1/B) = 0.23T^{-1}$, $0.073T^{-1}$ and 0.0149 T^{-1}. The number indicated by arrows are index, $N = B/\Delta(1/B)$, for each series of the oscillation. The amplitude of the oscillation was strongly dependent on temperature as shown in Fig. 3 where

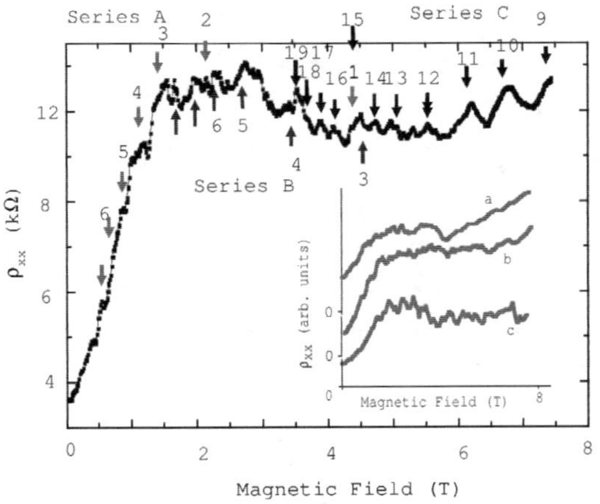

Figure 2. Magnetoresistance of monolayer graphene at the minimum conductivity point, V_g= -1.2V and T = 4.2 K. Arrows indicate the peaks of small oscillation which could be assigned to the oscillation periodic in $1/B$. Numbers indicate the indices of Landau levels. Inset shows magnetoresistance at the minimum conductivity point for different samples. Each sample showed different oscillatory behaviors. T = 4.2 K.

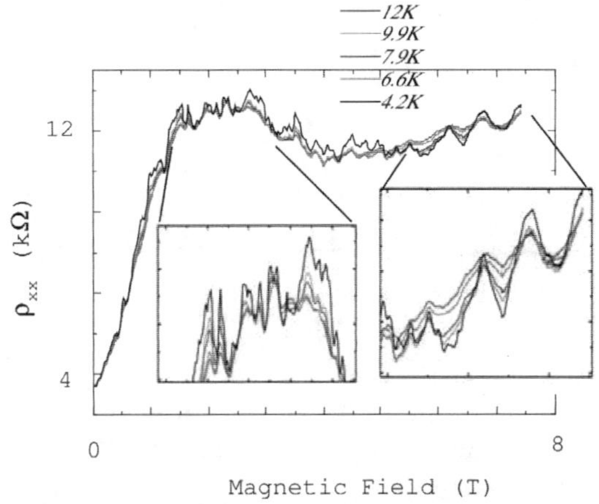

Figure 3. Temperature dependence of magnetoresistance of monolayer graphene at the minimum conductivity point, V_g= -1.2 V. Insets are enlargements of two different regimes in the main panel.

traces between T = 4.2K and 12 K is presented. The oscillation amplitude diminished with increasing temperature.

The periodicity in 1/B and the temperature dependence, are characteristic features of Shubnikov-de Haas (S-dH) oscillation. Ideally, only zero-mode Landau level is formed at the charge neutrality point. Therefore no resistance oscillation due to S-dH effect should appear here. However, in actual device S-dH oscillation is possible to appear because monolayer graphene is under strong influence of potential fluctuation because the conduction plane is exposed directly to atmosphere and/or to the substrate, which allows ionized impurities and/or contaminations to create local potential offsets. The influence of the potential fluctuation becomes increasingly significant in transport measurement as the gate voltage approaches the Dirac point. The fluctuation potential creates domains of electrons and holes, which is called electron-hole puddles[5]. Therefore there is a possibility that S-dH oscillations originating from some puddles were observed at the charge neutrality point.

Assuming that the oscillation stemmed from the S-dH effect, we estimated the carrier density n from the period of the S-dH oscillation. For Series A, B and C, n was estimated to be 3.9×10^{11} cm^2, 1.2×10^{12} cm^2 and 6.1×10^{12} cm^2, respectively. These variation in carrier densities could be caused by local shift of effective gate voltage. Variations of the effective gate voltage from the minimum conductivity point were estimated to be 5.9, 18.5, and 91V, using a relation $n = \alpha V_g/e$, where α is a specific capacitance of graphene and back-gate, which is 107 aF/μm^2 for our device. These results naturally led us to the possibility that the graphene sample should have at least three distinct domains biased locally.

We have estimated cyclotron mass m^* for the oscillation by fitting the amplitude of the oscillation to Lifsitz formula, $T/\sinh(2\pi^2 k_B m^* T/\hbar eB)$, where T is temperature and B is magnetic field. For each series of oscillations A, B and C, m^* was estimated to be $0.015m_0$,

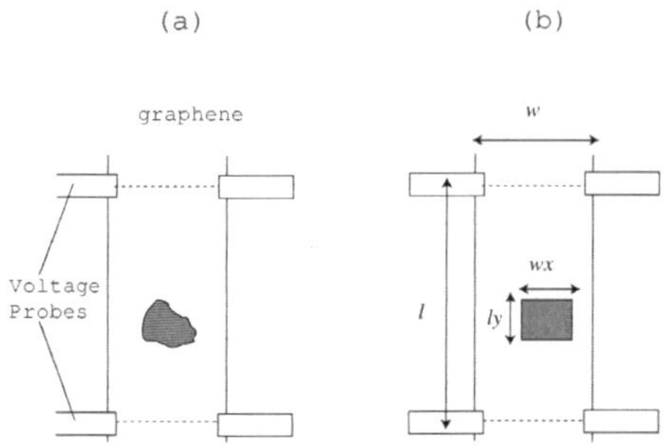

Figure 4. Illustration of (a) an example of a locally biased domain and (b) a model domain for resistance calculation. w and l are width and length. x and y denotes sizes of the domain relative to sample size.

$0.031m_0$, $0.07m_0$, respectively. The cyclotron masses are different because potential offsets, and therefore carrier densities are different. In monolayer graphene, cyclotron mass estimated using Lifsitz formula increases with increasing carrier density as $\propto \sqrt{n}$ [1]. From results in Ref. 1, cyclotron mass for each series of oscillation was estimated to be $0.015m_0$, $0.03m_0$, $0.06m_0$ for given carrier density, which agreed with our experiment fairly well.

Next we estimated domain size from the magnitude of the resistance oscillation. Above 2T, ρ_{xx} is about 13 kOhm, and magnitude of oscillation was about 1 kOhm. On the other hand, values of ρ_{xx} for the locally biased domains were estimated to be about 2.4 kOhm at V_g=6V and about 1.2 kOhm at V_g=20 V. The shape of the domain may be an isolated island in shape as illustrated in Fig. 4. Here we tentatively analyze for a simplified model where the island is a rectangular in shape as shown in Fig 4(b) because we had no information about the exact shape. We roughly calculated the variation of the entire resistance with respect to the change of the two probe conductance of the domain. Oscillation of 1 kOhm corresponded to a domain with width and length of about 1/10 of the sample size.

In transport measurements, information about potential fluctuation can be extracted from gate voltage dependence of zero-field conductivity using relations given in Ref. [5]. Typical value of the charge fluctuation at the minimum conductivity point can be estimated from the width of the dip in the graph of V_g dependence of σ_{xx}. This, however, gives magnitude of potential fluctuation averaged over the sample, not giving any information about size which is causing the fluctuation. Recently electron-hole puddles were visualized by a measurement of local potential using a scanning single electron transistor (SET) microscope [8]. They reported that sizes of the puddles range from 30 nm to sub-micron. The present estimation is qualitatively in good agreement with their estimation. Also, magnitude of the residual carrier density of the order of 10^{11} cm^2, is approximately the same except for the Series C, in which the value is larger by an order of magnitude for Series C.

4. Conclusion

Magnetoresistance oscillation could be observed in monolayer graphene at the minimum conductivity point. Major magnetoresistance fluctuations having been observed at the minimum conductivity point stemmed from the S-dH effect arising from potential fluctuation. From the oscillation, we could estimate magnitude of the potential relevant to the oscillation, and typical size of the domains responsible for the oscillation.

References

[1] Novoselov KS, Geim AK, Morozov SV, Jiang D, Katsnelson MI, Grigorieva IV, Dubonos SV and Firsov AA 2005 *Nature* **438** 197

[2] Shon NH and Ando T 1998 *J. Phys. Soc. Jpn.* **67** 2421

[3] Ando T 2007 *J. Phys. Soc. Jpn.* **75** 074716

[4] Checklesky JG, Li L and Ong NP 2008 *Phys. Rev. Lett.* **100** 206801

[5] Adam S, Hwang EH, Galitski VM, and Sarma SD 2007 *Proc. Natl. Acad. Sci. USA* **104** 18392

[6] Cho S and Fuhrer MS 2008 *Phys. Rev. B* **77** 081402

[7] Yagi R, Fukada S, Kobara H, Ogita N, and Udagawa M 2010 *Physica E* **42** 673

[8] Martin J, Akerman JN, Ulbricht G, Lohmann T, Smet JH, Kliting Kv, and Yacoby A 2008 *Nature Phys.* **4** 144

A study on the magnetic behavior of Nd-Fe-B/α-Fe nanocomposite films

S Madeswaran[1], S Tamano[2], S Goto[2] and K Tokiwa[1,2]

[1]Polyscale Technology Research Center, Tokyo University of Science
2641 Yamazaki, Noda, Chiba 278-8510, Japan

[2]Department of Applied Electronics, Tokyo University of Science, Japan
2641 Yamazaki, Noda, Chiba, 278-8510, Japan

E-mail: smades@rs.noda.tus.ac.jp

Abstract. Nanocomposite Nd-Fe-B / α-Fe thin films were prepared by sputtering successively a Nd-rich $Nd_{21}Fe_{64}B_{15}$ and a Fe targets in a multilayer structure using radio frequency (RF) magnetron gun. We have studied the influence of thickness of α-Fe layer on the magnetic properties of Nd-Fe-B / α-Fe thin films. A nanocomposite thin film with 15nm thick α-Fe layer prepared on 550°C heated substrate gives the highest energy product, 190 kJ/m^3 along with a coercivity of 950 kA/m. Magnetic hysteresis loop measurement shows that the hard (Nd-Fe-B) and soft (α-Fe) layers are exchange coupled firmly for the films deposited on heated substrate whereas the two layers are decoupled for room temperature deposited and post annealed films.

1. Introduction

Nanocomposite materials, consisting of a fine mixture of a magnetically soft phase with high magnetization exchange coupled to a high anisotropy hard phase have been widely studied due to their application as high energy product and low cost permanent magnets [1-7]. A vital efforts are being performed by several researchers to increase the energy product, BH_{max}, in permanent magnet in the form of bulk [1-2] as well as in thin film [5-7]. However the practical values of remanence and maximum energy product are much lower than the theoretical values predicted by micromagnetic calculations [3-4] for nanocomposite having ideal exchange coupling between the grains of two phases. The main reason for such low values is due to large difference between ideal and practical grain size and distribution, and especially crystalline orientation of hard phase. Films based on a tetragonal $Nd_2Fe_{14}B$ phase are obviously the potential candidates because it possesses superior hard magnetic properties such as high spontaneous magnetization and large coercivity [1-2, 5-7]. Thin films of Nd-Fe-B with α-Fe nanocomposite magnets prepared by sputtering deposition in tri-layer form [5] with the soft layer sandwiched between the two hard magnetic layers showed enhanced remanence. The understanding of such multilayer system in the thin film is of considerable interest since the properties of sample could be optimized by controlling the thickness of each layer. Furthermore, the Nd-rich phase in connection with $Nd_2Fe_{14}B$ phase leads to higher coercivity with preferred orientation [8-10]. We therefore investigated Nd-rich Nd-Fe-B multilayer with soft magnetic (α-Fe) layer combination.

2. Experiment

Thin films of layer structure, Ta (35nm) /NdFeB (120nm)/ α-Fe (x nm)/ NdFeB (120nm)/Ta(20nm) where x = 0, 15, 30 60 and 90, were deposited by sequential sputtering of Ta (50mm Ø x 5mm size), Nd-rich with a nominal composition of $Nd_{21}Fe_{65}B_{14}$ (50mm Ø x 2mm) and Fe (50mm Ø x 0.1mm) targets onto glass substrate in a multiple gun RF plasma chamber with a base pressure of ~$8x10^{-5}$ Pa. The $Nd_{21}Fe_{65}B_{14}$ target was prepared by arc melting in an Ar atmosphere and successive spark plasma sintering (sintering temperature 700°C and pressure 30kN) after well ground the ingot. The Ta buffer layer used here is to enable texturing of $Nd_2Fe_{14}B$ phase and also to prevent diffusion of substrate materials into the film [11]. A set of samples were prepared directly on heated substrate at 550°C and the samples will be referred to as type-I. Another set of multilayer were made by depositing hard (NdFeB) and soft (Fe) magnetic layer at room temperature while Ta buffer layer deposited at 550°C to maintain crystalline buffer phase with (110) orientation for consistency of growth interface of hard magnetic layer. After deposition at room temperature the samples were annealed at 650°C for 15 min in an Ar atmosphere (where the O_2 content is reduced to 10^{-22} torr with the help of oxygen controller, STLab SiOC-200DC, connected with annealing furnace) to crystallize both the phases. These post annealed samples will be referred to as type-II. For all experiment, the deposition conditions such as pressure of Ar (high purity) gas, sputter power and distance between target and substrate were sustained at 5 Pa, 130 W and 60mm, respectively. X-ray diffraction (XRD) analysis was performed with a Rigaku Ultima-III diffractometer utilizing Cu-Kα radiation. Microstructural investigations were performed with a JEOL (JSM-7001F) field emission scanning electron microscope (FE-SEM). Magnetic properties were measured in both perpendicular and parallel direction to the film plane by vibrating sample magnetometer (VSM) with a maximum field of 20 kOe.

3. Results and Discussion

The XRD patterns of type-I films with layer form Ta(35nm)/NdFeB(240nm)/Ta(20nm), Ta (35nm) /NdFeB (120nm)/ α-Fe (15 nm)/ NdFeB (120nm)/Ta(20nm) and Ta (35nm) /NdFeB (120nm)/ α-Fe (60 nm)/ NdFeB (120nm)/Ta(20nm) are shown in figure 1. For the films deposited with and without α-Fe layer, intensity peaks corresponding *(004), (006)* and *(008)* plane of crystalline $Nd_2Fe_{14}B$ phase occur with a strong *(00l)* texture, that is, c-axis tending to be perpendicular to the film plane. After introducing a 15nm thick α-Fe layer in hard magnetic layer like sandwich structure and increasing its thickness further, the XRD patterns show increasing trend of intensity of the peak presents at 44.6°, which can be attributed to main reflection of α-Fe phase superimpose with *(006)* peak of $Nd_2Fe_{14}B$, and indicating the increase of α-Fe phase in the film. It is also interesting to note that intensities of peaks appearing at 30.3° and 50.7° assigned for Nd-rich and Nd_xO_y phases are decreasing for increase of thickness of α-Fe layer. The decrease of Nd-rich phase is comparable to the speculation made by Kato et al [6] who specified that Nd-rich phase depends on volume fraction of Fe, and its relative concentration in the film decreases with increasing volume fraction of Fe. The Nd-rich phase usually consists of complex Nd containing intermetallic compound and Nd_xO_y [12-13]. Additionally, the x-ray results show a unknown intensifying peak (at 35.2° - marked as * in pattern) with respect to α-Fe layer thickness and this is presumably due to the presence of Fe_3B, $Nd_1Fe_4B_4$ and Fe_x Nd_y.

Figure 2 shows the XRD patterns of type-II films, Ta/NdFeB/α-Fe (x) /NdFeB/ Ta with x = 0, 15 and 30 nm. The annealing for these films was performed rapidly with a ramp rate of 150°C/Sec according to the reports [14-15] that showed enhanced magnetic properties for rapid thermal annealed NdFeB thin films compared with a long time anneal as the rapid thermal temperature rise leads to more nucleation centers. The x-ray diffraction analyses of these films show the evidence of $Nd_2Fe_{14}B$ phase at the same time the preferred c-axis orientation of Nd-Fe-B film is not vertical to the film plane since there are many peaks with exclusion of definite intensity of *(00l)* crystalline planes. It is believed that the Nd_xO_y phase might support the growth of $Nd_2Fe_{14}B$ phase with c-axis texture when the deposition temperature is high; however, it obstructs the growth of texturing $Nd_2Fe_{14}B$ phase at low deposition temperatures and subsequent annealing treatment [16-17].

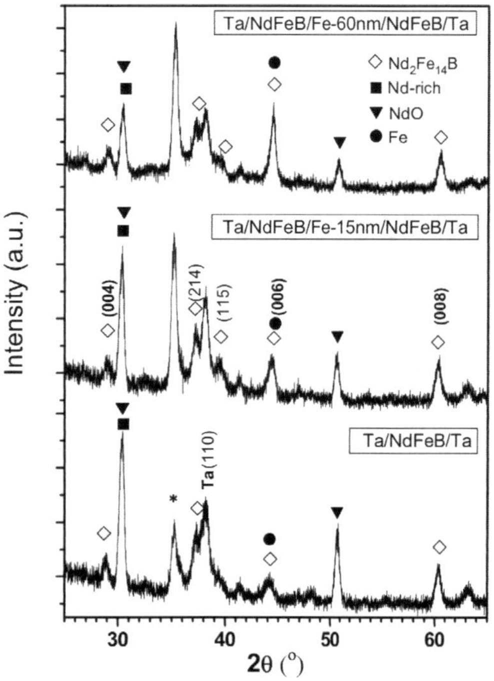

Figure 1. XRD patterns of the films deposited at 550°C.

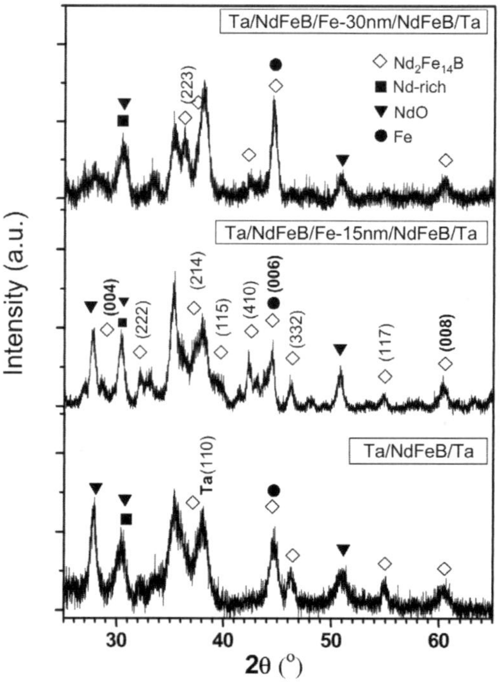

Figure 2. XRD patterns of the films deposited at room temperature and post annealed at 650°C.

To study about the grain growth, a film in the form of Ta/NdFeB(240nm)/Fe(30nm) was developed at high temperature (550°C) and analyzed by FE-SEM. Figure 3a shows FE- SEM image of this sample surface. The grains are irregular rectangular like shape and the average gain size is about 90 nm. Figure 3b displays a compositional mapping of the same area of (a). In this imaging mode, the gray contrast arises from the α-Fe phase, while the white contrasts arise from the $Nd_2Fe_{14}B$ phase. From this image the approximate grain size of α-Fe is estimated around 30 nm. Compositional mapping were also observed at cross sectional view of type-I and type-II films and are shown in figure 3c and d, respectively. The α-Fe layer or grains could not be seen in type-I sample whereas in the case type-II the image displays a clear thick gray line corresponding to the α-Fe phase. From the images 3b, c and d, it can be said that the α-Fe grains in type-I are embedded well in the Nd-Fe-B matrix compared with type-II film. FE-SEM imaging observation also infers that the type-I films are always having higher roughness compared with type-II because of higher atomic mobility at high temperature deposition.

The magnetic behavior at room temperature of type-I films are shown in the left column of figure 4. The solid and dashed lines represent measuring field direction along perpendicular (out-plane-$\perp$) and parallel (in-plane-$\|$) to the film plane, respectively. Demagnetization correction has not been performed for any of the measurement. The single phase like hysteresis loops for these samples measured along perpendicular direction have no two-step demagnetization process indicating that the soft and hard phases are exchange coupled. Moreover, the out-plane magnetization values are always larger than the in-plane, implying a good alignment of $Nd_2Fe_{14}B$ grains in c-axis which is consistent with x-ray diffraction data displayed in figure 1.

Figure 3. (a) FE-SEM image of Ta/NdFeB(240nm)/ α-Fe(30nm) film developed at 550°C, (b) compositional mapping of area (a), (c) Cross sectional view of Ta/NdFeB/α-Fe/NdFeB/Ta film of type-I and (d) type-II.

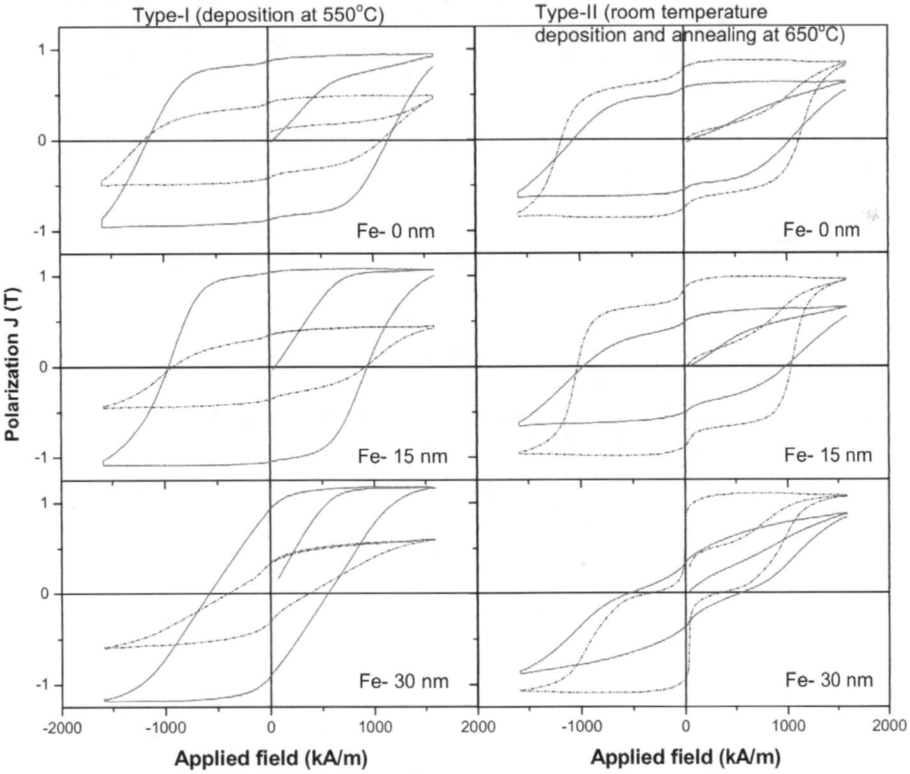

Figure 4. Room temperature hysteresis loops of type-I (left column) and type-II samples (right column) in the form of Ta(35nm)/NdFeB(120nm)/ α-Fe(x nm) /NdFeB(120nm) /Ta(20nm) with x =0, 15 and 30 (Solid curves : $H\perp$ film plane, dotted curves $H \parallel$ film plane).

The right column of figure 4 represents the magnetic properties of type-II samples. The NdFeB layer without α-Fe phase show a kink near zero applied field indicating the existence of secondary phase and this result coincides with the investigation made by Tokumaru et al [18]. For the addition of α-Fe layer, the two-step demagnetization process is pronouncing well and the step length drastically increases depending on α-Fe layer thickness. It obviously deduces that the soft and hard phases are decoupled, that is, there is poor exchange coupling between the phases presented in the film. Another point to note worth is that this type of samples exhibit larger in-plane magnetization values compared with out-plane. This is in agreement with X-ray patterns showed in figure 2 and suggesting that the easy magnetization axis is along the film plane.

Figure 5 summarizes the saturation magnetization (J_s), coercivity (H_c), maximum energy product (BH_{max}) and remanence ratio (J_r/J_s) of type-I sample as a function of α-Fe layer thickness. As described by many researchers [6, 13-14], who studied magnetic behavior against volume fraction of soft phase, it can obviously be seen that magnetization decreases and coercivity increases with increasing soft layer thickness. The J_s increases with the Fe content as one would expect since Fe has a larger moment (per gram) than NdFeB. The H_c decreases rapidly from 1150 kA/m to 950 kA/m for adding 15nm thick α-Fe layer since this composite system now has a smaller anisotropy per unit volume. These two opposition factors enable the maximum energy product to peak point at 15 nm thick α-Fe layer. The remanence ratio is higher for 15 nm and it is decreasing with further increase of thickness, indicating the deterioration of formation of $Nd_2Fe_{14}B$ crystallographic texturing. This deterioration may arise from the distribution of shape and size of grains. As a result, the maximum of BH_{max} value of 190 kJ/m^3 can be achieved for a nanocomposite film with 15nm α-Fe layer. However, the obtained maximum BH_{max} is smaller than the reported values of isotropic nanocomposite [2], this is mainly because of small magnetization value caused by the presence of non-magnetic phases as depicted by X-ray diffraction (figure 1). For the film of Ta/NdFeB(240nm)/Fe(30nm) layer which is subjected to compositional mapping (figure 3b), the magnetic behavior is more or less same as the tri-layer film in the form of Ta/NdFeB(120nm)/ α-Fe(30 nm)/NdFeB(120nm)/Ta.

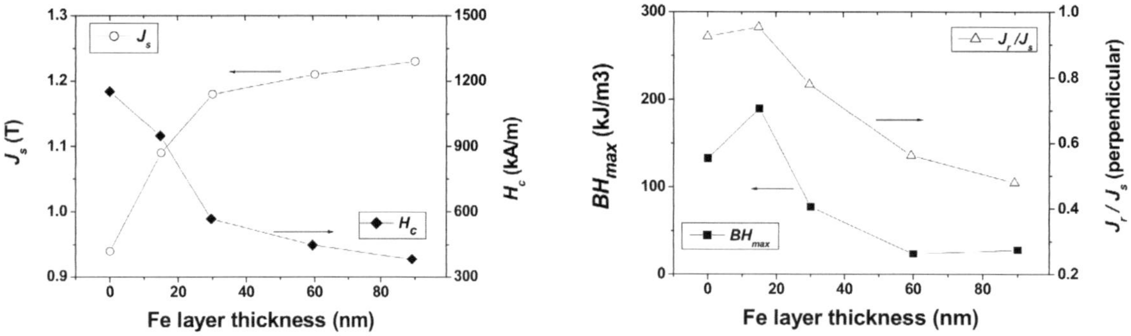

Figure 5. Magnetic properties of Ta(35nm)/NdFeB(120nm)/ α-Fe(x nm)/NdFeB(120nm)/Ta(20nm) (x =0, 15, 30, 60, 90) thin films (type-I) with different thickness of Fe layer.

4. Conclusion

The effect of α-Fe layer thickness on the magnetic properties and exchange coupling interaction of hard-soft-hard magnetic tri-layers of type-I (deposited at 550°C), and type-II (room temperature deposition and being annealed at 650°C for 15 min), were investigated. The results show that α-Fe layer is fully coupled to the Nd-Fe-B phase and nanocomposite structure can be developed for type-I films. The exchange coupled tri-layer with 15nm thick α-Fe gives a significant magnetic properties: H_c=950 kA/m, J_r = 1.04 T, $(BH)_{max}$ = 190 kJ/ m^3 (23.8 MGOe). If we increase the thickness of α-Fe layer to 30nm, the coercivity is greatly decreased to 570 kOe and texturing of the magnetic film also starts to deteriorate. Type-I films exhibit easy magnetization axis along perpendicular to film plane, where as type-II exhibit easy magnetization axis along film plane with decoupling between hard and soft phases.

Acknowledgement
This research work was supported by "Academic Frontier" Project for Private Universities: matching fund subsidy from MEXT (Ministry of Education, Culture, Sports, Science and Technology), 2006-2010.

References:

[1] Bauer J, Seeger M, Zern A, and Kronmuller H 1996 *J. Appl. Phys.* **80** 1667
[2] Liu S, Higgins A, Shin E, Bauser S, Chen C, Lee D, Shen Y and Huang M Q 2006 *IEEE Trans. Magn.* **42** 2912
[3] Kneller E F and Hawig R 1991 *IEEE Trans. Magn.* **27** 3588
[4] Skomski R and Coey J M D 1993 *Phys. Rev. B* **48** 15812
[5] Shindo M, Ishizone M, Kato H, Miyazaki T and Sakuma A 1996 *J. Magn. Magn. Mater.* **161** L1
[6] Kato H, Ishizone M, Koyama K and Miyazaki T 2005 *J. Magn. Magn. Mater.* **290-291** 1221
[7] Cui W B, Zheng S J, Liu W, Ma X L, Yang F, Yao Q, Zhao X G and Zhang Z D 2008 *J. Appl. Phys.* **104** 053903
[8] Makita K and Yamashita O 1999 *Appl. Phys. Lett.* **74** 2056
[9] Watanabe N, Umemoto H, Itakura M, Nishida M and Machida K 2009 *IOP Conf. Series: Mater. Sci. Eng.* **1** 012033
[10] Madeswaran S, Tokumaru R, Tamano S, Goto S, Tokiwa K and Watanabe T 2009 *J. Phys.: Conf. Series* **191** 012024
[11] Hannemann U, Fähler S, Oswald S, Holzpfel B and Schultz L 2002 *J. Magn. Magn. Mater.* **242** 1294
[12] Minowa T, Shimao M, Honshima M 1991 *J. Magn. Magn. Mater.* **97** 107
[13] Yue M, Zhang J X, Liu W Q and Wang G P 2004 *J. Magn. Magn. Mater.* **271** 364
[14] Jiang H and O'Shea N J 2000 *J. Magn. Magn. Mater.* **212** 59
[15] Kojima A, Makino A and Inoue A 2000 *J. Appl. Phys.* **87** 6576
[16] Chen S L, Zheng J G, Liu W and Zhang Z D 2007 *J. Phys. D: Appl. Phys.* **40** 1816
[17] Serrona L K E B, Sugimura A, Adachi N, Okuda T, Ohsato H, Sakamoto I, Nakanishi A, Motokawa M, Ping D H and Hono K 2003 *Appl. Phys. Lett.* **82** 1751
[18] Tokumaru R, Tamano S, Goto S, Madeswaran S, Tokiwa K and Watanabe T 2009 *J. Phys.: Conf. Series* **191** 012021

Effect of current annealing on electronic properties of multilayer graphene

S Tanaka[1,2,4], **H Goto**[1,4], **H Tomori**[1,2,4], **Y Ootuka**[2], **K Tsukagoshi**[3,4], **A Kanda**[1,4]

[1]Institute of Physics, University of Tsukuba, Tsukuba 305-8571, Japan
[2]Tsukuba Research Center for Interdisciplinary Materials Science (TIMS), University of Tsukuba, Tsukuba 305-8571, Japan
[3]MANA, NIMS, Namiki, Tsukuba, Ibaraki 305-0047, Japan
[4]CREST-JST, Kawaguchi 332-0012, Japan

E-mail: `tanaka@lt.px.tsukuba.ac.jp`

Abstract. While ideal graphene has high mobility due to the relativistic nature of carriers, it is known that the carrier transport in actual graphene samples is dominated by the influence of scattering from charged impurities, which almost conceals the intrinsic splendid properties of this novel material. The common techniques to improve the graphene mobility include the annealing in hydrogen atmosphere and the local annealing by imposing a large biasing current. Although annealing is quite important technique for the experimental study of graphene, detailed evaluation of the annealing effect is lacking at present. In this paper, we study the effect of the current annealing in multilayer graphene devices quantitatively by investigating the change in the mobility and the carrier density at the charge neutrality point. We find that the current annealing sometimes causes degradation of the transport properties.

1. Introduction

Graphene, a single atomic layer of graphite, is expected to take important roles both in fundamental physics and in industrial applications. In fundamental physics, single layer graphene (SLG) attracts a lot of attentions, because an electron in SLG behaves as a massless relativistic particle, called a Dirac fermion, due to the linear energy dispersion. On the other hand, graphene is expected to be applied to future nanoelectronic devices due to its possible high mobility, reaching 200,000 cm^2/Vs at room temperature.[1]

In spite of such expectations, the mobility achieved in actual graphene devices has been limited to one-tenth of the intrinsic value. Possible scattering factors behind the decrease in the mobility include (1) phonons in graphene itself and in the SiO_2 substrate, (2) the structural defects and ripples in the graphene, and (3) the charged impurities originated from the contamination of residual resist, adsorbed gas molecules, and the SiO_2 substrate. We can expect that the influence of factor (2) is relatively small, since it is known that graphene films obtained with the commonly-used micromechanical cleavage technique[2] has a nearly perfect crystal structure over several micrometer range.[3] On the other hand, factor (3) seems to be quite serious since by using a scanning single electron transistor (SET) microscope, Martin *et al.* observed spatial fluctuations of the carrier density, called electron and hole puddles, even at the Dirac point, at which no carrier is supposed to be present.[4] Besides, theoretical study [5] has shown that the charged impurities is the origin of the linearly increasing conductance as a function of the gate voltage, which is commonly observed in experiments.[6] Thus, the charged impurities dominates the transport properties in graphene at low temperatures, concealing the intrinsic relativistic physics.

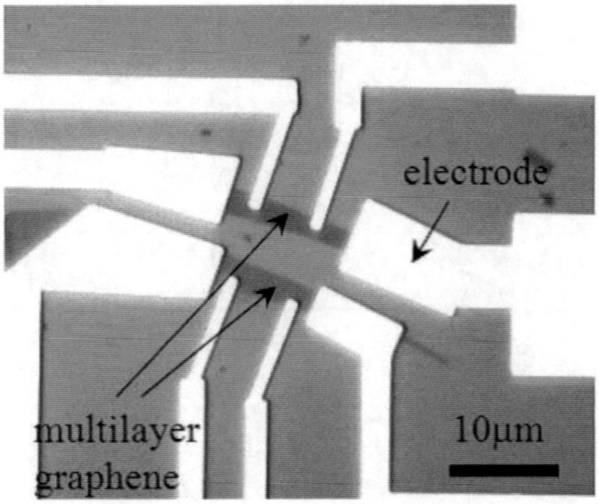

Figure 1. Optical microscope image of a sample.

Annealing is generally known as a method to remove such charged impurities. Two types of annealing methods are used commonly. In the first, graphene on a Si/SiO_2 substrate is annealed around 300 °C in Ar/H_2 atmosphere. The second is the so-called current annealing by applying a large biasing current to graphene at low temperatures.[7] Although these methods are widely used to remove contamination and adsorbed molecules, how they works has not been investigated in detail so far. In this paper, we study the effect of the current annealing using two techniques: evaluation of the carrier density by the transport measurement and observation of the graphene surface by an atomic force microscope (AFM).

2. Results and discussion

Multilayer graphene (MLG) films were obtained by the standard micromechanical cleavage technique of Kish graphite.[2] Four current and voltage leads are connected to an MLG film by using the electron beam lithography and the deposition of Cr/Au. The highly-doped Si/SiO_2 substrate was used as a back gate. Figure 1 shows an optical micrograph of a sample. The samples were set in a low temperature probe stations, and cooled to 4 - 10 K. Current-voltage characteristics were measured as a function of the gate voltage while the temperature rose to room temperature. Conductance was calculated from the current-voltage characteristics. After the first measurement finished, we cooled the sample again without taking it out of the vacuum chamber. Current annealing was performed with a current density around $1 \times 10^8 \mathrm{A/cm^2}$ for 10 - 20 minutes just before the second measurement started (4 K).

Here, we compare electrical transport before and after the current annealing. Figures 2(a) and 2(b) show the conductance of sample 1 as a function of the gate voltage, $\sigma(V_{\mathrm{gate}})$ for several temperatures before and after the current annealing, respectively. $\sigma(V_{\mathrm{gate}})$ takes a minimum value σ_{min} at a gate voltage corresponding to the charge neutrality point. From the linear part of $\sigma(V_{\mathrm{gate}})$ seen at large $|V_{\mathrm{gate}}|$, we calculated the field effect mobility μ. By using the Drude formula and the relation between the carrier density and the gate voltage, μ is given by,

$$\mu = \frac{1}{\alpha e}\frac{\mathrm{d}\sigma}{\mathrm{d}V_{\mathrm{gate}}}, \tag{1}$$

where $\alpha = 7.2 \times 10^{10}/\mathrm{cm^2V}$ is the induced carrier density by the gate voltage change of 1 V. Note that μ is defined both for electrons and holes. The minimum carrier density n_0 corresponding to the charge neutrality point is calculated from the minimum conductivity σ_{min}, as follows:

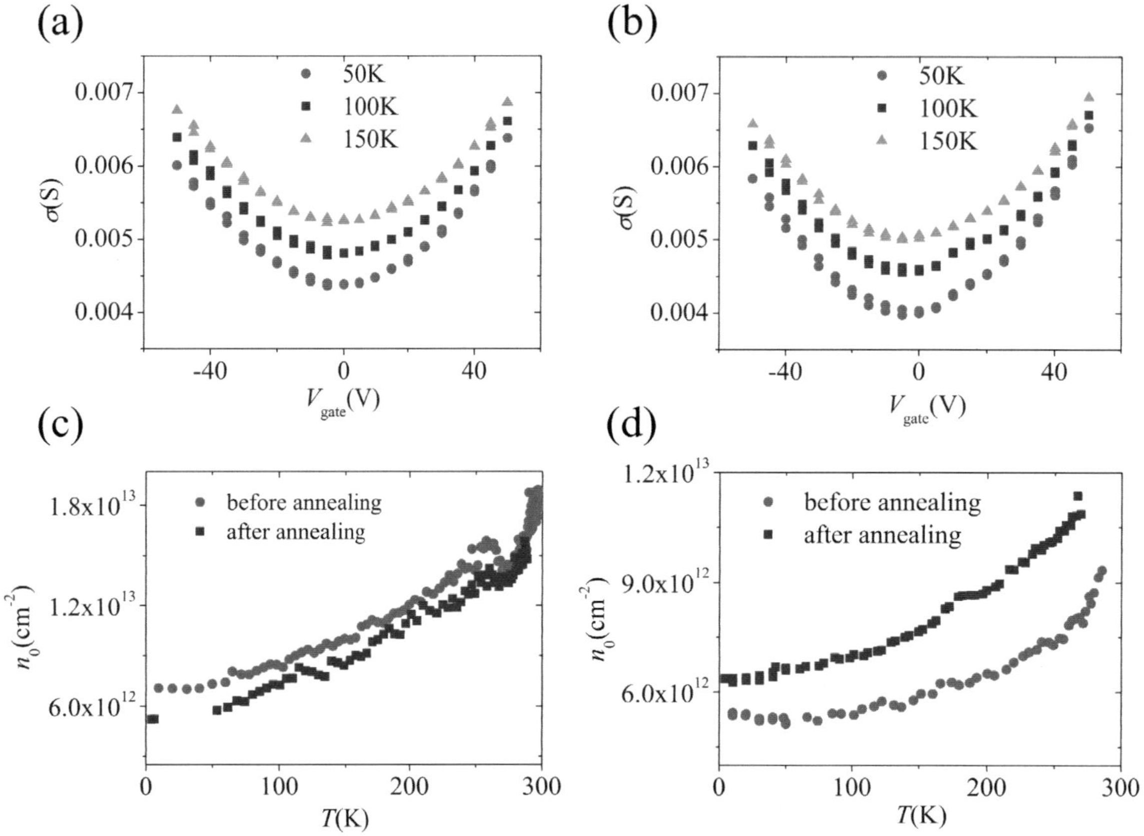

Figure 2. Gate voltage dependence of the conductivity before (a) and after (b) annealing in sample 1. Minimum carrier density as a function of temperature before and after the current annealing for sample 1(c) and sample 2(d).

$$n_0 = \frac{\sigma_{\min}}{e} \cdot \frac{2}{(\mu_h + \mu_e)}, \tag{2}$$

where μ_h and μ_e are the hole and electron mobilities, respectively.

Figure 2(c) shows the minimum carrier density $n(0)$ before and after the annealing for sample 1. By the current annealing, the mobility rises and the minimum carrier density decreases for all temperature range; the hole mobility changes from 3,500 to 4,300cm^2/Vs, and the carrier density from 6.9×10^{12} to 5.2×10^{12}/cm^2 at 4 K. This result suggests that the number of charged impurities were decreased by the current annealing. Sample 2, however, shows the opposite result: the mobility decreases and the minimum carrier density increases by the current annealing, as shown in Figure 2(d): the mobility changes from 8,500 to 5,300cm^2/Vs and the carrier density changes from 5.3×10^{12} to 6.3×10^{12}/cm^2 at 4 K. This result indicates the increase of the charged impurities despite of the current annealing. So far we have examined seven samples; four samples behaved like sample 1: the current annealing increased the mobility and decreased the minimum carrier density. On the other hand, three samples were similar to sample 2: the current annealing decreased the mobility and increased the minimum carrier density.

To investigate the origin of the contradicting effects of the current annealing on the transport properties of graphene, we observed the surface of the samples after the current annealing with the AFM. The results are shown in Fig. 3. Sample 1 (Fig. 3(a)) has a flat surface with asperity below 1

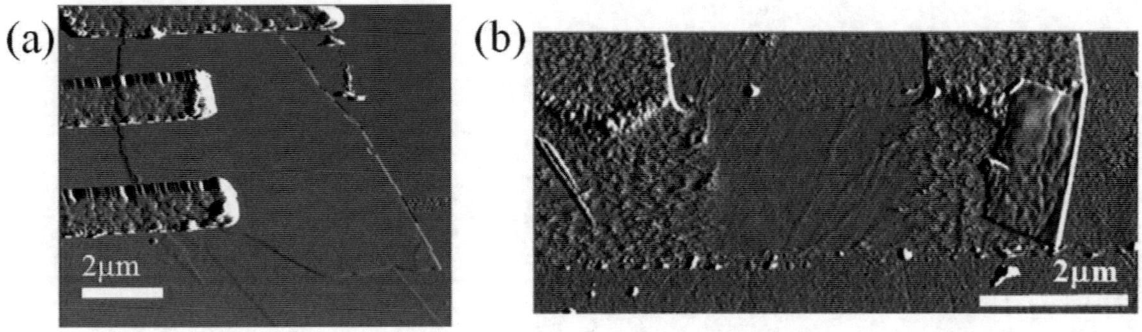

Figure 3. AFM image of sample 1 (a) and sample 2 (b) after annealing.

nm. In contrast, sample 2 has rough surface near the electrodes with roughness from several nanometers to several 10 nm. At present, the origin of this rough surface is not clear, but we have confirmed that the formation of rough surface is closely related to the current annealing and that the degradation of the mobility and the increase of the minimum carrier density are seen only in samples with rough surfaces induced by current annealing. The details will be published elsewhere.

3. Conclusion

We have studied the effect of current annealing in graphene by transport measurements and AFM observation. The results are classified into two categories; in the first, the current annealing removes contamination as expected, and increases the mobility and decreased the carrier density by the order of $10^{12}/\mathrm{cm}^2$. On the contrary, in the second case, the current annealing accumulates the contamination, suppresses the mobility and increases the carrier density. To improve the transport properties of graphene using the current annealing, we need to optimize the annealing conditions.

Acknowledgments

This work is partly supported by the Grants-in-Aid for Scientific Research (17069004) from the Ministry of Education, Culture, Sports, Science and Technology (MEXT) of Japan, and the Inter-university Cooperative Research Program of the Institute of Materials Research, Tohoku University.

References

[1] Chen J H, Jang C, Xiao S, Ishigami M and Fuhrer, 2008 *Nat. Nanotech* **3** 206
[2] Novoselov K S, Geim A K, Morozov S V, Jiang D, Zhang Y, Dubonos S V, Grigorieva I. V, Firsov A A, 2004 *Science* **306** 666
[3] Stolyarova E, Rim K, Ryu S, Maultzsch J Kim, P, Brus L E, Heinz T F, Hybertsen M S and Flynn G W, 2007 *Natl. Acad. Sci.* **104** 9209
[4] Martin J, Akerman N, Ulbricht G, Lohmann T, Smet J H, von Klitzingand K and Yacoby A, 2008 *Nat. Phys.* **4** 144
[5] Adam S, Hwang E H, Galitski V M and Das Sarma S, 2007 *Proc. Nat. Acad. Sci.* **104** 18392
[6] Novoselov K S, Geim A K, Morozov S V, Jiang D, Katsnelson M I, Grigorieva I V, Dubonos S V, Firsov A A, 2005 *Nature* **438** 197
[7] Moser J, Barreiro A and Bachtold A, 2007 *Appl. Phys. Lett.* **91** 163513

Nickel(0)-catalyzed [3 + 3] cyclodimerization of methylenecyclopropanes

T Taniguchi [1] **M Ohashi** [1,2,4] **S Saito** [3] **and S Ogoshi** [1,4]

[1] Department of Applied Chemistry, Graduate School of Engineering, Osaka University, Suita, Osaka 565-0871, Japan
[2] Center for Atomic and Molecular Technologies, Osaka University, Suita, Osaka 565-0871, Japan
[3] Department of Chemistry, Faculty of Science, Tokyo University of Science, Shinjuku, Tokyo 162-8601, Japan

ohashi@chem.eng.osaka-u.ac.jp, ogoshi@chem.eng.osaka-u.ac.jp

Abstract. Treatment of ethyl cyclopropylideneacetate (ECPA) with a catalytic amount of $Ni(cod)_2$ and tricyclohexylphosphine at room temperature underwent dimerization reaction to give a 6-membered ring compound and a 5-membered ring compound in the ratio of 3:7. After extensive optimization, it was found that slow addition of ECPA by syringe drive over 5 h at 60 °C afforded the 6-membered product selectively.

1. Introduction

Transition-metal-catalyzed cyclodimerization or cycloaddition reactions of methylenecyclopropanes (MCPs) have provided an atom-efficient, powerful strategy for the synthesis of a variety of cyclic compounds [1-3]. Ni(0) is also known to promote [2 + 2] and [3 + 2] cyclodimerization reactions of MCPs to give cyclobutane and cyclopentane derivatives, respectively (Scheme 1) [4-10]. Recently, we demonstrated that nickel(0) complexes bearing a strong σ-donor ligand promote oxidative addition of the cyclopropane ring adjacent to a carbonyl group to yield a six-membered oxanickelacycles [11-21]. We, therefore, investigated a stoichiometric reaction of $Ni(cod)_2$ (cod = 1,5-cyclooctadiene) with ethyl cyclopropylideneacetate (**1**; ECPA) in the presence of PCy_3, (Cy = c-C_6H_{11}) leading to the selective formation of a 1,2-bis-exoalkylidenecyclohexane framework as a result of a [3 + 3] cyclodimerization reaction [8; 22-36]. We also developed a Ni-catalyzed regio- and diastereo-selective [3 + 3] cyclodimerization of electron-deficient alkylidenecyclopropanes.

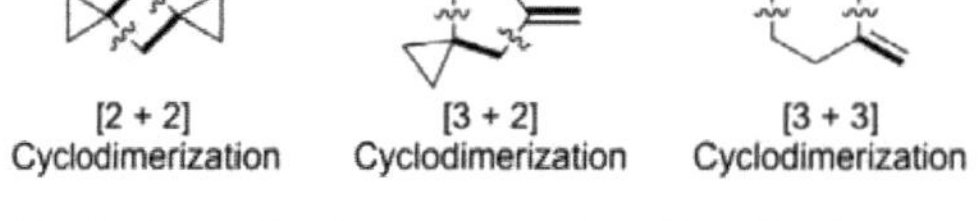

Scheme 1. Cyclodimerization products of methylenecyclopropane.

---

[4] To whom any correspondence should be addressed.

2. Results and Discussion

Treatment of **1** with $Ni(cod)_2$ in the presence of 2 equiv of PCy_3 at -15 °C resulted in the quantitative formation of an η^2-ECPA complex (**2**) (Scheme 2). Monitoring the further reaction at room temperature by means of NMR spectroscopy, gradual decomposition of **2** was observed while no other reaction intermediates were detectable. **2** was consumed within 60 h to yield a 1:1 mixture of $Ni(cod)_2$ and an unexpected Ni(0) complex (**3**), in which the coordinating 1,2-bis-exoalkylidenecyclohexane framework arose from the [3 + 3] cyclodimerization of ECPA via selective cleavage of the proximal C-C bond *trans* to the ethoxycarbonyl group.

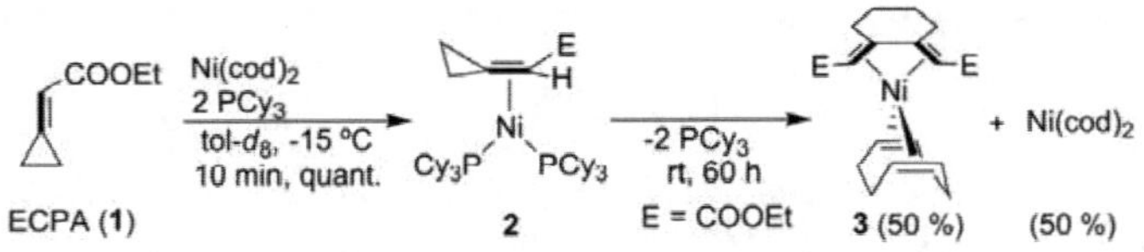

Scheme 2. Reaction of **1** with $Ni(cod)_2$ in the presence of PCy_3

In contrast, the reaction with 1-cyclopropylidene-2-propanone (**4**) proceeded in a different manner. Stoichiometric treatment of $Ni(cod)_2$ and PCy_3 with 2 equiv of **4** gave a Ni(II) complex (**5**) quantitatively (Scheme 3).

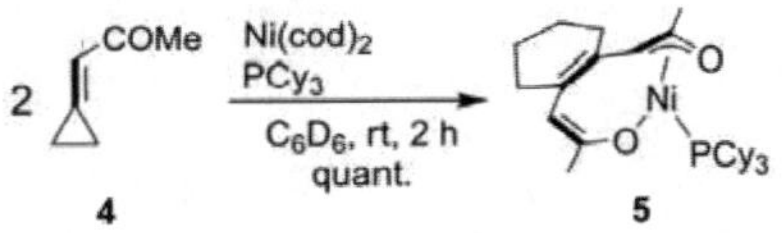

Scheme 3. Reaction of **1a** with $Ni(cod)_2$ in the presence of PCy_3

A catalytic application of the selective [3 + 3] cyclodimerization of **1** was examined. The reaction of **1** with a catalytic amount of $Ni(cod)_2$ (10 mol%) and PCy_3 (20 mol%) was conducted at room temperature. However, the expected [3 + 3] cycloaddition product (**6**) was obtained in 38% yield and a [3 + 2] cycloaddition product (**7**) was formed as a major product (Table 1, run 1). Elevating the reaction temperature (60 °C) improved the yield of **6** to 81% (run 2). Slow addition of **1** to the toluene solution of a $Ni(cod)_2$ and PCy_3 mixture over 5 h resulted in the selective formation of **6** to 90 % yield, and the major product, (*E*,*E*)-**6**, was isolated in 70% yield (run 3).

Table 1. $Ni(0)/PCy_3$-Mediated [3 + 3] Cyclodimerization of **1**[a].

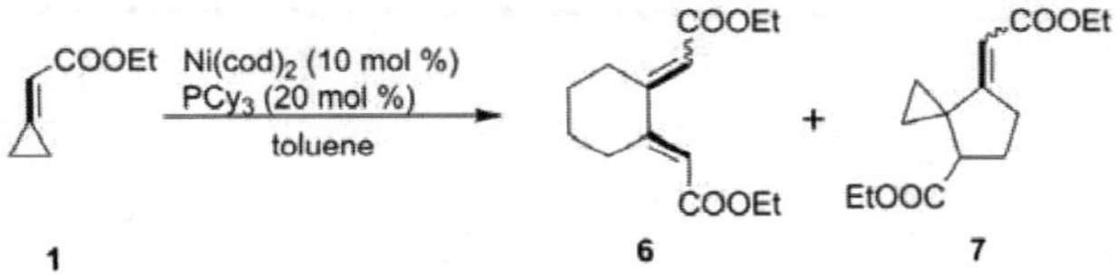

run	Conditions	Yield of **6** (%)[b] [*EE*:*EZ*]	Yield of **7** (%)[b] [*EE*:*EZ*]
1	rt, 7 h	38 [95:5]	62 [100:0]
2	60 °C, 1 h	81 [87:13]	19 [75:25]
3	60 °C, 6 h[c]	90(70) [92:8]	10 [20:80]

[a] General conditions: [**1**] = 0.2 M.

[b] Determined by ^1H NMR analysis of the crude reaction mixture. Cited yields in parentheses are of isolated major regioisomer.

[c] After terminating slow addition of **1** (for 5 h), the reaction mixture was further stirred for 1 h.

3. Conclusions

In summary, we have demonstrated the unique nickel-catalyzed [3 + 3] cyclodimerization of alkylidenecyclopropanes to yield 1,2-bis-exoalkylidenecyclohexanes.

4. Experimental Details

Reaction of 1a with $Ni(cod)_2$ in the presence of PCy_3: $Ni(cod)_2$ (5.5 mg, 0.020 mmol) and PCy_3 (11.2 mg, 0.040 mmol) were dissolved in 0.3 mL of toluene-d_8, and the solution was cooled down to -15 °C. Another reaction vial was charged with **1** (2.4 mg, 0.019 mmol) and toluene-d_8 (0.3 mL), and the solution of **1** was also pre-cooled down to -15 °C. These solutions were combined at -15 °C and the mixture was kept at the temperature for 10 min. NMR analysis showed that the quantitative formation of an η^2-ECPA complex (**2**) was observed. Monitoring the further reaction at room temperature by means of NMR spectroscopy, the gradual decomposition of **2** was observed to yield an unexpected Ni(0) complex (**3**). Spectral data for **3**: ^{1}H NMR (400 MHz, C_6D_6, rt, δ/ppm): 1.00–1.10 (m, 2H, -C*H*H-CH_2-C(=C)-), 1.17 (t, *J* = 7.0 Hz, 6H, -OCH_2CH_3), 1.24–1.34 (m, 2H, -CH*H*-CH_2-C(=C)-), 1.74–1.83 (m, 2H, *cod*), 1.85–1.94 (m, 2H, *cod*), 2.20 (s, 2H, C=C*H*-COO), 2.23–2.34 (m, 2H, *cod*), 2.33–2.45 (m, 2H, *cod*), 2.38 (m, 2H, -CH*H*-C(=C)-), 3.13 (ddd, *J* = 18.4, 7.2, 5.8 Hz, 2H, -C*H*H-C(=C)-), 3.76 (m, 2H, *cod*), 4.12 (dq, *J* = 10.8, 7.0, Hz, 2H, -OC*H*H-CH_3), 4.30 (dq, *J* = 10.8, 7.0, Hz, 2H, -OCH*H*-CH_3), 4.91 (m, 2H, *cod*). ^{13}C NMR (100 MHz, C_6D_6, rt, δ/ppm): 15.0 (q, J_{CH} = 126.3 Hz, -OCH_2CH_3), 22.2 (t, J_{CH} = 126.7 Hz, -CH_2-CH_2-C(=C)-), 27.1 (t, J_{CH} = 127.8 Hz, -CH_2-C(=C)-), 29.6 (t, J_{CH} = 122.1 Hz, *cod*), 30.7 (t, J_{CH} = 127.0 Hz, *cod*), 59.1 (t, J_{CH} = 145.9 Hz, -OCH_2CH_3), 59.9 (d, J_{CH} = 161.0 Hz, =*C*H-COO), 91.8 (d, J_{CH} = 157.3 Hz, *cod*), 92.4 (d, J_{CH} = 153.9 Hz, *cod*), 108.3 (s, *C*=CH-COO), 171.0 (s, =*C*O). Anal. Calcd for $C_{22}H_{32}NiO_4$: C, 63.04; H, 7.69. Found: C, 63.08; H, 7.90.

Reaction of 4 with $Ni(cod)_2$ in the presence of PCy_3: $Ni(cod)_2$ (11.1 mg, 0.04 mmol) and PCy_3 (11.2 mg, 0.04 mmol) was dissolved in 0.5 mL of C_6D_6. **4** (7.7 mg, 0.08 mmol) was added to the solution at room temperature. The reaction mixture was observed to change from yellow to dark-brown. After 2 h, NMR analysis showed that the full consumption of **4** and the quantitative formation of **5** were observed. Spectral data for **5**: ^{1}H NMR (400 MHz, C_6D_6, rt, δ/ppm): 1.02–1.32 (m, 12H, *Cy*), 1.48–2.05 (m, 21H, *Cy*), 1.75 (obscured by PCy_3, 2H, -CH_2-), 1.88 (s, 3H, η^3-CH_3CO-), 1.95 (obscured by PCy_3, 2H, -CH_2-), 1.98 (s, 3H, η^1-CH_3CO-), 2.34 (br s, 2H, η^3-CH_3COCH=C-CH_2-), 2.65–2.75 (m, 2H, η^1-CH_3COCH=C-CH_2-), 4.75 (br s, 1H, η^3-CH_3COC*H*-), 5.11 (s, 1H, η^1-CH_3COC*H*-). ^{13}C{^{1}H} NMR (100 MHz, C_6D_6, rt, δ/ppm): 24.0 (s, η^3-CH_3CO-), 24.2 (s, -CH_2-), 24.4 (s, -CH_2-), 26.3 (s, η^1-CH_3COCH=C-CH_2-), 26.8 (4-Cy), 27.9 (d, J_{PC} = 5.0 Hz, 3-Cy), 28.0 (d, J_{PC} = 5.0 Hz, 3-Cy), 28.6 (s, η^1-CH_3CO-), 29.9 (d, J_{PC} = 4.0 Hz, 2-Cy), 32.2 (d, J_{PC} = 18.1 Hz, P-*C*H-), 34.5 (s, η^1-CH_3COCH=C-CH_2-), 91.5 (d, J_{PC} = 12.0 Hz, η^3-CH_3C(O)-*C*H-), 100.8 (η^1-CH_3C(O)=*C*H-), 113.4 (s, η^1-CH_3COCH=*C*-CH_2-), 135.9 (s, η^3-CH_3COCH=*C*-CH_2-), 152.0 (s, η^3-CH_3*C*O-), 166.7 (s, η^1-CH_3*C*O-). ^{31}P{^{1}H} NMR (109 MHz, C_6D_6, rt, δ/ppm): 37.3 (s). Anal. Calcd for $C_{30}H_{49}NiO_2P$: C, 67.81; H, 9.29. Found: C, 67.66; H, 9.04.

Ni(0)/PCy_3-mediated [3 + 3] cyclodimerization of 1: $Ni(cod)_2$ (27.6 mg, 0.10 mmol) and PCy_3 (57.0 mg, 0.20 mmol) were dissolved into toluene (10 mL). To the solution was added dropwise a toluene solution (5.0 mL) of **1** (126 mg, 1.0 mmol) at 60 °C over 5 h. The reaction mixture was observed to change from yellow to deep-red. After stirred for further 1 h, the reaction mixture was concentrated in *vacuo*. The resulting crude oil was purified by filtration through a silica pad (Et_2O as an eluting solvent) to give a mixture of **6** (83% for (*E,E*); 7% for (*E,Z*)) and **7** (2% for (*E*); 8% for (*Z*)). This mixture was further purified by HPLC ($CHCl_3$) to afforded 88.1 mg of (*E,E*)-**6** (70%, isolated yield) as colorless oil. Spectral data for (*E,E*)-**6**: ^{1}H NMR (400 MHz, C_6D_6, rt, δ/ppm): 0.99 (t, *J* = 7.0 Hz, 6H, -OCH_2CH_3), 1.36 (m, 4H, -CH_2CH_2C=CH), 3.01 (m, 4H, -CH_2CH_2C=CH), 3.99 (q, *J* = 7.0 Hz, 4H, -

OCH_2CH$_3$), 5.85 (t, J = 1.8 Hz, 2H, -C=CH-COO). $^{13}C\{^1H\}$ NMR (100 MHz, C_6D_6, rt, δ/ppm): 14.3 (-OCH$_2$$CH_3$), 25.9 (-$CH_2CH_2$C=CH), 30.2 (-CH$_2$$CH_2$C=CH), 59.8 (-O$CH_2CH_3$), 115.3 (C=$C$HCOO), 160.4 ($C$=CHCOO), 165.9 (=$C$O). HRMS (EI) Calcd for $C_{14}H_{20}O_4$ 252.1362, Found m/z 252.1353. The minor products, (*E*,*Z*)-**6**, and (*E*)- and (*Z*)-**7**, were characterized based on the comparison of the chemical shifts of the vinyl proton (δ_H 5.50 (t) and 5.98 (s) for (*E*,*Z*)-**6**; 5.25 (t, 2.0 Hz) for (*E*)-**7**; 5.78 (s) for (*Z*)-**7**) with those reported values [8]. Yields and *E*/*Z* ratio were determined by NMR analysis of the crude reaction product.

References

[1] Binger P and Büch H M 1987 *Top. Curr. Chem.* **135** 77
[2] Dzhemilev U M, Khusnutdinov R I and Tolstikov G A 1991 *J. Organomet. Chem.* **409** 15
[3] Binger P and Schmidt T 1997 *Methoden Org. Chem.* (Houben-Weyl) E17c 2217
[4] Binger P 1972 *Angew. Chem. Int. Ed.* **11** 309
[5] Binger P and McMeeking J 1973 *Angew. Chem.* **85** 1053
[6] Binger P 1973 *Synthesis* 427
[7] Binger P, Doyle M J and Benn R 1983 *Chem. Ber.* **116** 1
[8] Kawasaki T, Saito S and Yamamoto Y 2002 *J. Org, Chem.* **67** 4911
[9] Tamaru Y *Modern Organonickel Chemistry* Wiley-VCH: Weinheim Germany, 2005.
[10] Noyori R, Kumagai Y, Umeda I and Talaya H 1972 *J. Am. Chem. Soc.* **94** 4018
[11] Ogoshi S, Nagata M and Kurosawa H 2006 *J. Am. Chem. Soc.* **128** 5350
[12] Tamaki T, Nagata M, Ohashi M and Ogoshi S 2009 *Chem. Eur. J.* **15** 10083
[13] The six-membered oxanickelacycle was found to be a key intermediate in Ni(0)-catalyzed cyclodimerization of cyclopropyl phenyl ketone, see also Ref [14; 15]
[14] Liu L and Montgomery J 2006 *J. Am. Chem. Soc.* **128** 5348
[15] Lloyd-Jones G C 2006 *Angew. Chem. Int. Ed.* **45** 6788
[16] Ogoshi S, Oka M-a and Kurosawa H 2004 *J. Am. Chem. Soc.* **126** 11082
[17] Ogoshi S, Ueta M, Arai T and Kurosawa H 2005 *J. Am. Chem. Soc.* **127** 12810
[18] Ogoshi S, Tonomori K.-i, Oka M.-a and Kurosawa H 2006 *J. Am. Chem. Soc.* **128** 7077
[19] Ogoshi S, Ikeda H and Kurosawa H 2007 *Angew. Chem. Int. Ed.* **46** 4930
[20] Ogoshi S, Arai T, Ohashi M and Kurosawa H 2008 *Chem. Commun.* 1347
[21] Ohashi M, Kishizaki O, Ikeda H and Ogoshi S 2009 *J. Am. Chem. Soc.* **131** 9160
[22] Nugent W A and Calabrese J C 1984 *J. Am. Chem. Soc.* **106** 6422
[23] Negishi E, Cederbaum F E and Takahashi T 1986 *Tetrahedron Lett.* **27** 2829
[24] Nugent W A, Thorn D L and Harlow R L 1987 *J. Am. Chem. Soc.* **109** 2788
[25] Negishi E, Holmes S J, Tour J M, Miller J A, Cederbaum F E, Swanson D R and Takahashi T 1989 *J. Am. Chem. Soc.* **111** 3336
[26] Urabe H and Sato F 1996 *J. Org. Chem.* **61** 6756
[27] Tamao K, Kobayashi K and Ito Y 1989 *J. Am. Chem. Soc.* **111** 6478
[28] Onozawa S, Hatanaka Y and Tanaka M 1997 *Chem. Commun.* 1229
[29] Onozawa S and Hatanaka Y, Choi N and Tanaka M 1997 *Organometallics* **16** 5389
[30] Suginome M, Matsuda T and Ito Y 1998 *Organometallics* **17** 5233
[31] Uno T, Wakayanagi S, Sonoda Y and Yamamoto K 2003 *Synlett* 1997
[32] Miura T, Yamauchi M and Murakami M 2007 *Synlett* 2029
[33] Bhatarah P and Smith E H 1992 *J. Chem. Soc., Perkin Trans.1* **114** 2163
[34] Piers E, McEachern E J and Romero M. A 1997 *J. Org. Chem.***62** 6034
[35] Lomberget T, Bouyssi D and Balme G 2005 *Synthesis* 311
[36] Gandon V, Aubert C, Malacria M and Vollhardt K P C 2008 *Chem. Commun* 1599
[37] Binger P, Müller P, Podubrin S, Albus S and Krüger C 2002 *J. Organomet. Chem.***656** 288

The Use of Laminar Inorganic Salts to Make Organic Molecules Display New Properties at the Supramolecular Level in the Solid State

E Brunet[1], H M H Alhendawi, M Alonso, C Cerro, L Jiménez, O Juanes, M J Mata, A Salvador, M Victoria, E Rodríguez-Payán and J C Rodríguez-Ubis
Department of Organic Chemistry, Faculty of Sciences
Universidad Autónoma de Madrid, 28049-Madrid (Spain)

E-mail: ernesto.brunet@uam.es

Abstract. The design of porous solids of controlled molecular geometry for umpteen applications is a challenge of enormous technological and scientific importance. The placing of organic molecules between the layers of certain inorganic salts leads to enduring solid materials where the confinement makes the organic molecules change their properties or even display new ones at the supramolecular level. Past and ongoing results of our research group concerning the chemistry of metal phosphates/phosphonates are reviewed in relation with recognition, chemically-driven porosity changes, chiral memory and supramolecular chirality, luminescence signaling, photoinduced electron-transfer, gas storage and drug confinement.

1. Introduction

One of the goals of the 4th International Symposium on Atomic Technology (Kobe, November 2009) is to offer solutions to the still unresolved Feynman's controversial query already stated some fifty years ago [1]: "*What would the properties of materials be if we could really arrange the atoms (molecules) the way we want them?* The main objective of our research group is to provide a chemical, conceptual and practical rationale to humbly defy Feynman's challenge.

In general, the design of supramolecular materials is aimed along those lines and the literature contains many smart processes to achieve that, such as sol-gel chemistry [2], self-assembly [3], film making (Langmuir-Blodgett, etc.) [4], and the use of classical porous materials like zeolites, among many other approaches [5]. Yet, Feynman also conjectured "*What could we do with layered structures with just the right layers?* Along the years we have found that zirconium phosphate, a versatile layered inorganic salt, acts as a carving board where organic compounds can be covalently attached by low-temperature hydrothermal reactions to form 3D structures with properties and potential applications only limited by the imagination of the researcher. In particular, our project's objective is the *preparation of nanostructured materials with predictable structure*, based on organic-inorganic supramolecular lattices, insoluble and thermally stable, with specific functions like *i*) recognition; *ii*) chemically-driven porosity changes; *iii*) chiral memory and supramolecular chirality; *iv*) luminescence signaling; *v*) photoinduced electron-transfer; *vi*) gas storage; and *vii*) drug confinement. In this paper we shall present a brief overview of our findings in these fields.

2. Tools for the construction: Zirconium Phosphate (ZrP)

Figure 1 shows the two most common allotropic forms of this layered inorganic salt [6]. In the *alpha* form one may discover a single kind of phosphates bonded thru three oxygen atoms to three different

---

[1] To whom any correspondence should be addressed.

metals, each of the latter being hexa-coordinated. Every phosphate points an acidic OH group towards the interlayer region. In turn, the *gamma* phase contains two different kinds of phosphates, one internal (green tetrahedrons) using all oxygen atoms to bond four different Zr, and another kind (blue tetrahedrons) placed on the upper and lower external surfaces, bonded to only two Zr atoms and directing two acidic OH groups towards the interlayer space. The salt ZrP is very versatile [7], especially the γ-phase, because the surface phosphates may be topotactically replaced (i.e. preserving the layered structure) by other phosphorous functions like organic phosphonates of any kind: rigid or flexible, polar or non-polar, chiral or non-chiral or with any conceivable chemical function (Figure 2). In the case of using diphosphonates, the final structure ends up as a pillared 3D arrangement of inorganic lamellae covalently bonded to each other by appropriate organic moieties. The topotactic exchange may be multiple, either simultaneously performed by mixing different phosphorous derivatives or in sequence, by adding one after another. We thus have to our disposal a set of conceptual and practical tools to designing organic-inorganic materials of any imaginable nano-scaled structure whose arrangement can be predicted with quite reasonable accuracy.

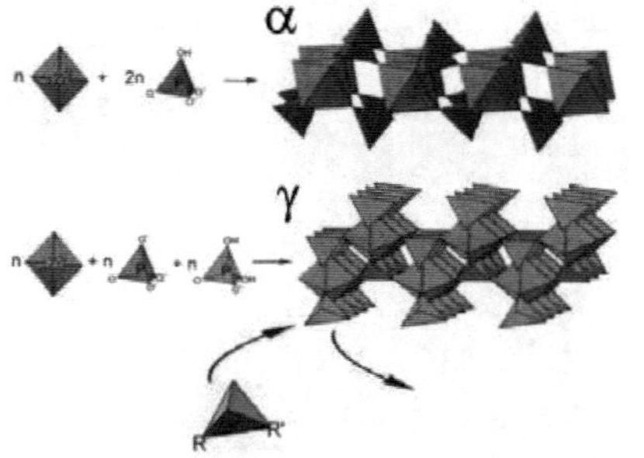

Figure 1. Most common allotropic forms of ZrP. In the γ-ZrP sketch the possible replacement of the surface phosphates by other phosphorous functions is schematically shown.

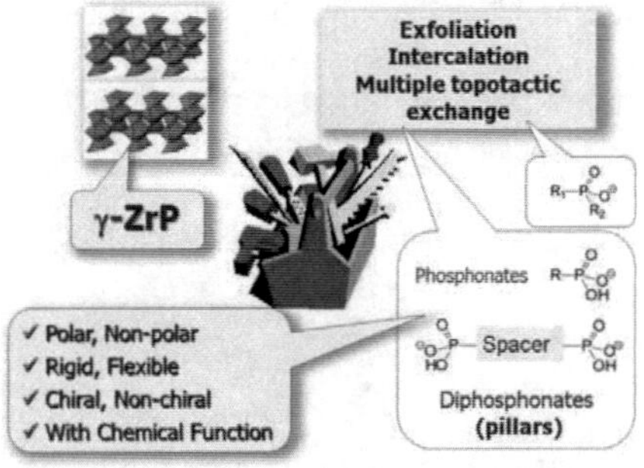

Figure 2. Schematic view of our conceptual and practical tools for the construction of organic-inorganic materials.

3. Recognition

Various crown ethers of different sizes, containing one or two phosphonate groups, where exchanged within the layers of γ-ZrP [8]. The resulting set of materials was treated with methylphosphonate to further exchange the remaining surface phosphate groups. Competitive experiments (Figure 3) showed that the recognition abilities towards alkaline ions, exerted in the solid-liquid interface by the confined crowns, were very different to those commonly exhibited in solution. Furthermore, the second phosphate/methylphosphonate exchange completely altered the crowns' response to the ions, thus showing that the nature of the whole matrix structure is crucial in the overall recognition process.

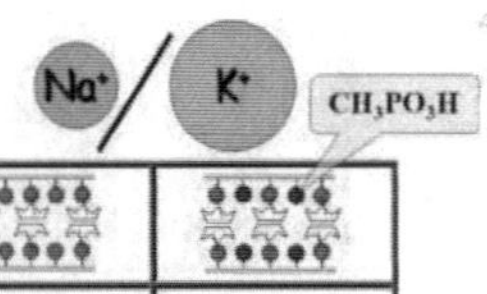

Material		
ZrP	0.9	-
12C4	6.6	2.0
15C5	0.7	-
18C6	0.3	1.2
Kriptofix	0.5	3.3

Figure 3. Summary of the results in competitive experiments for Na^+/K^+ recongnition.

4. Chemically-driven porosity changes

Diphosphonates derived from polyethylenglycol chains of different lengths (from di- to hexa-ethylenglycol) were prepared and exchanged at 25% level into γ-ZrP [9]. The corresponding pillared materials (Figure 4) were thus formed, bearing the general molecular formula of $Zr(PO_4)(H_2PO_4)_{0.75}[HO_3P(CH_2)_2(OCH_2CH_2O)_n(CH_2)_2PO_3H]_{0.125}$ (n = 2 to 6). The remaining surface phosphonates $[(H_2PO_4)_{0.75}]$ were quantitatively replaced by hypophosphite $[(H_2PO_2)_{0.75}]$ leading to two sets of materials, polar/polar (polar columns and polar surface phosphates) and polar/non-polar (polar columns and non-polar surface hypophosphites). Figure 5 summarizes the observed variation of interlayer distance (as measured by DRX) when the materials containing pentaethylenglycol diphosphonate (n = 5), as a representative example, were treated with methylamine in aqueous dispersion. It may be seen that the polar/polar material steadily augmented its interlayer distance with the increasing amount of intercalating amine. When no amine is present, the polar columns interact by hydrogen bonding with the surface phosphates and the layers are compressed to one another. The amine reacts with the surface phosphates and progressively disrupts the web of hydrogen bonding interactions, thus making the columns to stand up. The overall porosity of the material is thus heavily increased by the mild acid-base reaction. In the case of the polar/non-polar material the increase of interlayer distance occurs all of a sudden, within a very narrow pH range. The replacement of the surface phosphates by hypophosphite greatly diminishes the overall acidity of the material, the only acidic OH groups being those of the phosphonates. In the absence of amine, the polar columns still can establish O…H-P hydrogen bonds with the surface hypophosphite groups. When sufficient amount of amine is present, the few remaining acidic OH groups are quickly neutralized and the methylammonium ions act as wedges that pull the columns to suddenly rise. Therefore, the interlayer distance is doubled and the porosity is on the whole profoundly changed. To the best of our knowledge, this odd behaviour in the solid state, responding to a simple acid-base reaction in the solid-liquid interface, has never been observed before [10].

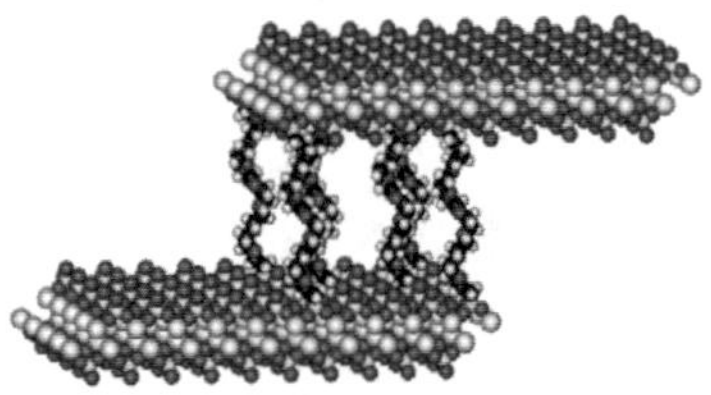

Figure 4. Animated molecular model of γ-ZrP exchanged with pentaethylenglycol diphosphonate with variable interlayer distance (see text).

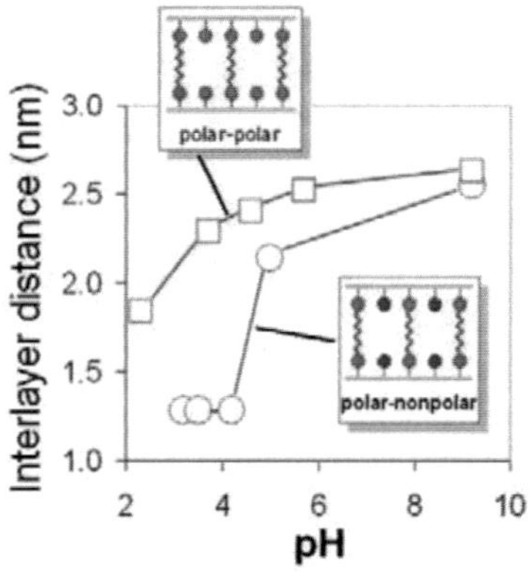

Figure 5. Interlayer distance variation with pH of the indicated materials (see text).

5. Supramolecular chirality and chiral memory

When we performed molecular models of the materials described in the previous section, we realized that to fit the largest achieved interlayer distance, the polyethylenoxa columns had to be helicoidally arranged. This is not surprising since polyethylenglycols are known to attain that conformation in aqueous solution. Yet, what makes a notable difference is that in our materials, the polyethylenoxa columns are covalently attached by both ends to the inorganic layers, thus conferring them a supramolecular order. Therefore, assuming the necessary conditions are established, the P/M helicity of the chains could be somewhat directed to render a certain degree of homochirality which could be detected by simple optical rotation measurements. Figure 6 outlines the experiments aimed to that observation. Dispersions of 5 mg in 10 mL of 1:1 water/acetone of either native γ-ZrP or exchanged at 25% with hexaethylenglycol diphosphonate (n = 6; *cf.* molecular formula in previous section) displayed no sizeable optical rotation. However, when optically pure (+)-phenethylamine was intercalated in the material containing hexaethylenglycol, a relatively large value of optical rotation was measured. We then smoothly replaced the enantiomerically pure amine by hexylamine and the optical rotation kept showing through. In absence of any (+)-phenethylamine, as shown by pertinent NMR experiments, the only source of optical activity should be the homochirality of the polyethylenoxa columns that was arranged by the concourse of the chiral amine and was maintained when the latter was replaced by the achiral one. This is an expression of supramolecular chiral memory; no similar cases can be found in the literature [11].

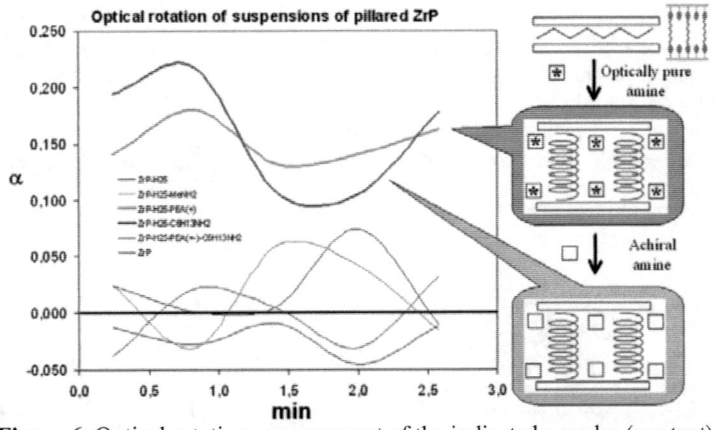

Figure 6. Optical rotation measurement of the indicated samples (see text).

We wonder what would happen if the chiral influence were intrinsic to the polyethylenoxa chains. To this end, we prepared enantiomerically pure diphosphonates derived from the addition of *R*-glycidol to diethyl vinylphosphonate and subsequent epoxide aperture with hexaethylenglycol (Scheme 1). The topotactic exchange of this diphosphonate into γ-ZrP was performed at 25% level and the material was positively characterized using the usual techniques. It occurred to us that the conformation of the polyethylenoxa chains attached to the inorganic layers could be changed if amines of appropriate length were intercalated. Figure 7 summarizes the results. As expected, interlayer distance steadily increased with the length of the amine. Molecular modelling showed that the organic chains are continuously changing their conformation from compressed random coiling at the shortest layer separation to stretched zigzag arrangement at the largest basal spacing. Optical rotation measured in 1:1 water/acetone dispersion passed through a maximum value when hexylamine was intercalated and the resulting interlayer distance allowed for the columns to be in their most ordered helicity, as the molecular models strongly suggested.

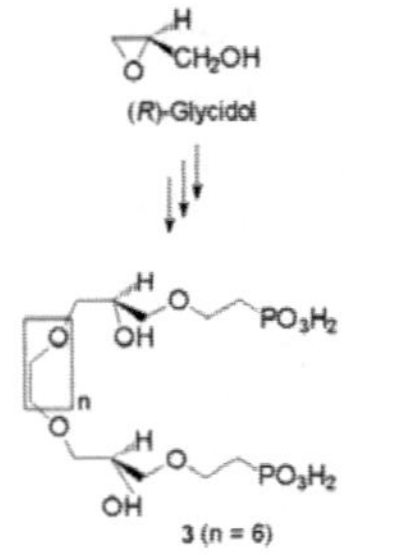

Scheme 1

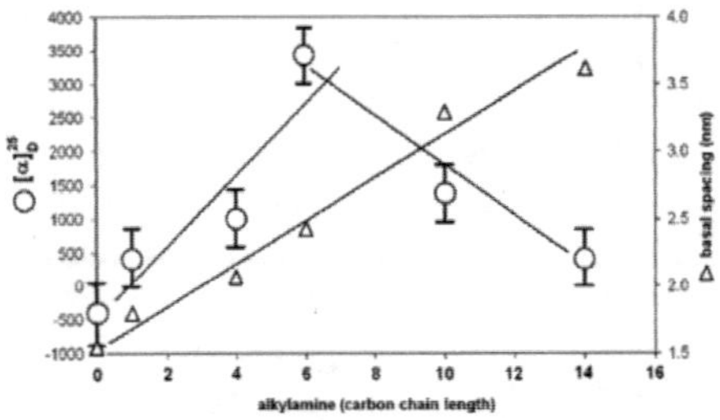

Figure 7. Variation of interlayer distance and optical rotation of γ-ZrP exchanged with the compound of Scheme 1 with the intercalation of alkylamines of increasing length (see text).

In our opinion this experiment clearly proved that chirality can be created at the supramolecular level in the solid state and finely tuned by mild reactions in the solid-liquid interface. There are no examples in the literature achieving this effect in such a clear-cut manner [12].

6. Luminescence signalling

The building of molecular systems able to display the strong luminescence of lanthanide metals is very active a field in chemistry [13]. To accomplish it two stringent conditions have to be met: *i*) a suitable organic chromophore should absorb light and efficiently transfer energy to the metal (*antenna effect*); *ii*) the coordination sphere of the metal should be free of water molecules because the OH oscillators easily quench metal emission. The white powder of γ-ZrP does not absorb visible light and remains reasonably transparent in the UV region. On the other hand, the pillared materials described in the previous section with polyethylenoxa columns may constitute an excellent spider web to enshroud the oxygenphylic lanthanide metals and isolate them within the solid matrix. Acid-base reactions with the surface phosphates may be taken as the driving force to intercalate suitable chromophores in the organic-inorganic lattice. Figure 8 schematizes part of the achieved results.

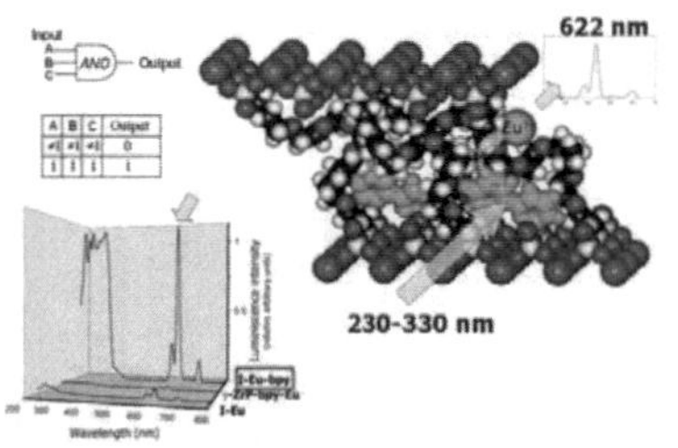

Figure 8. The lanthanide emission was only observed when ethylenpolyoxa columns, the 2,2'-bipiridyl and the metal were confined within γ-ZrP.

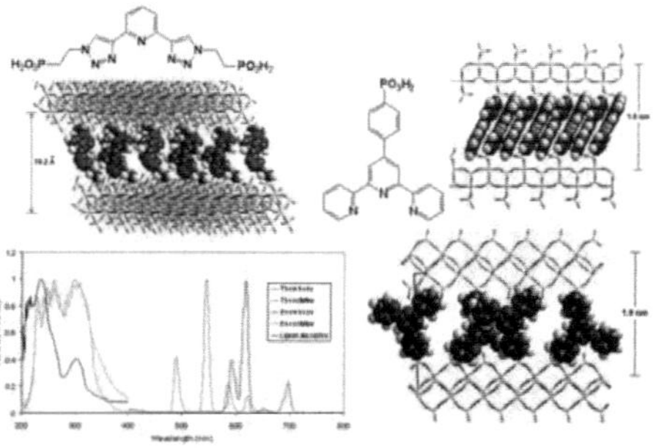

Figure 9. Exchanged γ-ZrP with chromophores derived from bis-triazolylpyridine and phenylterpyridine, showing the characteristic emissión of Eu^{3+} and Tb^{3+}.

In electronics terms the composite behaves as a triple AND gate because, for the metal luminescence to take place, it is necessary the gathering of the three components: lanthanide metal, polyethylenoxa columns and 2,2'-bipyridyl as sensitizer. These materials may find a number of different applications either in solid or in the solid-liquid interface. More sophisticated chromophores are being tried (Figure 9) and preliminary results showed the pursued strong luminescence of the metals [14].

7. Photoinduced electron transfer

There is no fundamental principle running against the possibility of attaining the proper arrangement of chemical components on a surface to accomplish long-lived photoinduced electron transfer [15], the first step in either the achievement of artificial photosynthesis or the construction of efficient solar cells. Inorganic γ-ZrP may constitute an excellent carving board to realize that goal. We have attached to it phosphonate derivatives of the $Ru^{II}(bpy)_3$ complex, well known by its excellent light-absorption and electron-donor properties, and placed side-by-side to it different electron donor acceptors ranging from relatively simple viologens to more elaborated fullerene derivatives. Figure 10 contains some molecular models of these complex structures which were positively characterized by the usual techniques [16].

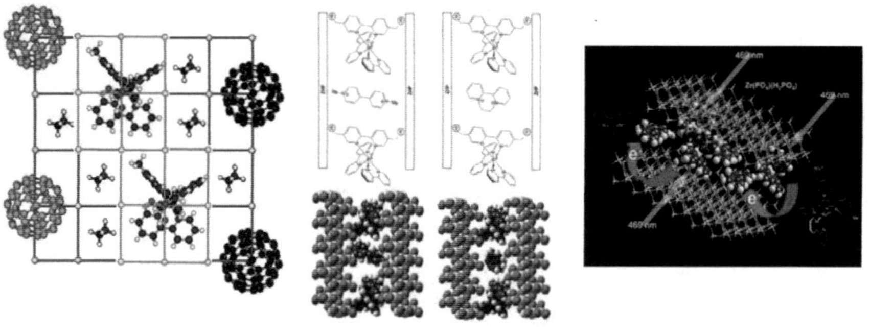

Figure 10. Idealized arrangement of electron-donor $Ru^{II}(bpy)_3$ and electron aceptors (fullerene derivatives or viologens) on the simplified surface of γ-ZrP.

The first important observation was that the phosphorescence emission of the $Ru^{II}(bpy)_3$ complex was heavily quenched by the presence of the electron-acceptor species, suggesting that the pursued electron transfer took place. Very recently, flash photolysis revealed some of these materials to have outstanding properties in that the separation of charges is long lived and the initially thought inert inorganic layer resulted semiconducting thus driving the separated electrons and/or holes along the material [17]. Additional results showed that these powders can be arranged as solar cells with promising efficiency [18].

8. Gas storage

Hydrogen is a very appealing energy vector - the release of its energy does not involve the noxious carbon dioxide. Yet, environmental-friendly production and safe storage/transportation methods must be devised if hydrogen is to be efficiently used as the clean energy carrier of the future. Although there are already many reasonably useful technical approaches, neither of the two problems is nowadays at a level of resolution which would make the use of hydrogen routinely possible. Concerning hydrogen storage, it may be attained among other procedures by *physisorption* in porous matrices [19]. We conjectured whether appropriate matrices based on γ-ZrP could serve the latter purpose. The thorough revision of the literature allows one to find numerous organic, inorganic and hybrid systems in which the storage of hydrogen has been tested, from the delusive carbon nanotubes to the very cleverly designed systems with almost unbelievable specific surface areas higher than 5000 m^2/g. The analysis of the different available structures points to the conclusion that a large accessible volume is desirable but by means of micropores or ultramicropores and therefore presenting a large internal contact surface to the elusive hydrogen molecules. Some interesting theoretical studies and experimental ones on zeolites also point to the benefit of having polarizing centers, namely small cations as for instance Li^+. Also the presence of transition metals with open coordination sites appears to be important. We believe that our set of tools (Figure 2) might allow us to design appropriate materials for hydrogen storage. Figure 11 summarizes our rationale and the best achieved results [20].

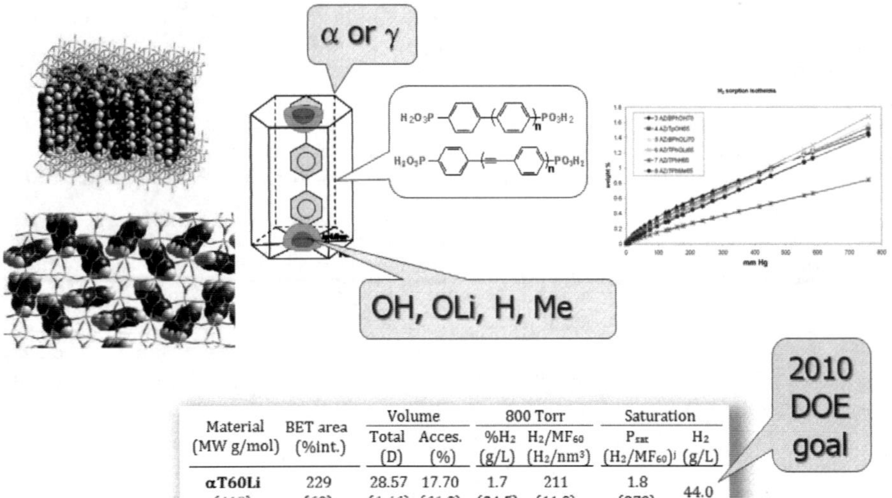

Material (MW g/mol)	BET area (%int.)	Volume: Total (D)	Volume: Acces. (%)	800 Torr: %H_2 (g/L)	800 Torr: H_2/MF_{60} (H_2/nm^3)	Saturation: P_{sat} (H_2/MF_{60})^j	Saturation: H_2 (g/L)
αT60Li (419)	229 (69)	28.57 (1.46)	17.70 (61.9)	1.7 (24.5)	211 (11.9)	1.8 (379)	44.0

Figure 11. Phosphonate-exchanged ZrP materials for hydrogen storage (see text).

The attachment of polyphenyl or polyphenylethynyl diphosphonates to either α- or γ-ZrP led to materials with slit-like ultramicropores of different length. Further exchange reactions led to polar or non-polar groups at both ends of the pores. This set of materials allowed us to check for a large number of different arrangements. The best results were obtained with the material named **αT60Li**, α-ZrP with 60% of terphenyldiphosphonates, the remainder phosphates bearing Li^+ as counterions. At 800 Torr and 77K, 1.7% w/w of hydrogen could be stored and the Department of Energy goal for 2010 (45 g of hydrogen per L of material) can be attained at 1.8 atm at 77K (Figure 11).

9. Drug confinement

Confinement of drugs into molecular matrices might be a good solution to their slow release at particular locations on the body [21]. We have preliminary studies where it can be shown that the erythromycin family antibiotics can intercalate into γ-ZrP, the driving force being the acid-base interaction between the amine group(s) of the organic compound and the acidic surface phosphates. The measurement of the interlayer distance *vs.* the stoichiometry of the reactants (Figure 12) shows that mono- and bilayers of erythomycin derivatives may be enclosed between the inorganic lamellae. More detailed studies will follow.

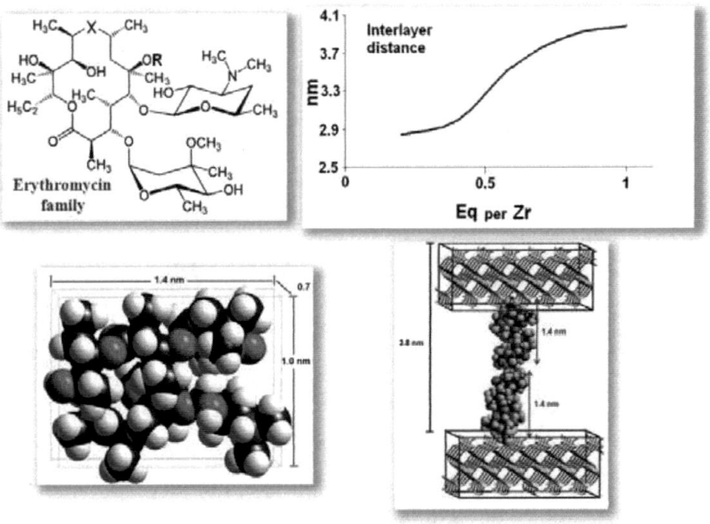

Figure 12. Intercalation of antibiotics of the erythromycin family (X = CO, R = H, erythromycin A; R = Me, clarithromycin; X = MeN-CH_2, R = H, azithromycin). The interlayer distance indicates the formation of mono- and bilayers of antibiotic between the inorganic lamellae at low and high stoichiometric ratio of the reactants, respectively. The x-ray crystal structure of clarithromycin and an ideal model of the bilayer are presented.

10. Conclusion

It has been our aim to show, within the limited length of this communication, the enormous possibilities that the building of organic-inorganic 3D structures does have in numerous fields of either basic or applied research. In particular, the use of ZrP allows the construction of rather complex composites where the confinement of the organic molecules confers them new and unexpected properties at the supramolecular level.

Acknowledgement

This work has been performed under the indirect support of ERCROS-Farmacia S.A. to which we are highly indebted. Unfortunately, we do regret that the financial backing from the Ministry of Science and Education of Spain has just ceased despite the generous grants received in the near past (MAT2002-03243, MAT2006-00570) and the more than reasonable results so far achieved.

References

[1] Feynman R 1960 *Eng. Sci.*, 22-36.

[2] Walcarius A and Collinson M M 2009 *Ann. Rev. Anal. Chem.* **2** 121-143.

[3] Hunter C A and Anderson H L 2009 *Angew. Chem., Int. Ed.* **48** 7488-7499; Kharlampieva E, Kozlovskaya V and Sukhishvili S A 2009 *Adv. Mater.* **21** 3053-3065; Liu Y and Yan H 2009 *Science* **325** 685-686.

[4] Yang J and Yan D 2009 *Chem. Soc. Rev.* **38** 2634-2645.

[5] Farrusseng D, Aguado S and Pinel C 2009 *Angew. Chem., Int. Ed.* **48** 7502-7513.

[6] Alberti G, Murcia-Mascaros S and Vivani R 1994 *Mat. Sci. Forum* 152-153; Vivani R, Alberti G, Costantino F and Nocchetti M 2007 *Microp. Mesop. Mat.* **107** 58-70.

[7] Alberti G, Costantino U, Dionigi C, Murcia-Mascaros S and Vivani R 1995 *Supram. Chem.* **6** 29-40; Clearfield A and Costantino U 1996 *Comprehensive Supramolecular Chemistry* **7** 107-149; Olivera-Pastor P, Maireles-Torres P, Rodriguez-Castellon E, Jimenez-Lopez A, Cassagneau T, Jones D J and Roziere 1996 *J Chem. Mat.* **8** 1758-1769.

[8] Brunet E, Huelva M and Rodriguez-Ubis J C 1994 *Tetrahedron Lett.* **35** 8697-8700; Brunet E, Huelva M, Vazquez R, Juanes O and Rodriguez-Ubis, J C 1996 *Chem. Eur. J.* **2** 1578-1584.

[9] Alberti G, Brunet E, Dionigi C, Juanes O, Mata M J, Rodriguez-Ubis J C and Vivani R 1999 *Angew. Chem., Int. Ed.* **38** 3351-3353.

[10] Brunet E, Mata M J, Juanes O and Rodriguez-Ubis J C 2004 *Angew. Chem., Int. Ed.* **43** 619-621; Brunet E, Mata M J, Alhendawi H M H, Cerro C, Alonso M, Juanes O and Rodriguez-Ubis J C 2005 *Chem. Mat.* **17** 1424-1433.

[11] Brunet E 2002 *Chirality* **14** 135-143.

[12] Brunet E, Mata M J, Juanes O, Alhendawi, H M H, Cerro C and Rodriguez-Ubis J C 2006 *Tetrahedron: Asym.* **17** 347-354; Brunet E, Juanes O and Rodriguez-Ubis J C 2009 *J. Mex. Chem. Soc.* **53** 154-161.

[13] Brunet E, Juanes O and Rodriguez-Ubis J C 2007 *Cur. Chem. Biol.* **1** 11-39.

[14] Brunet E, Mata M J, Juanes O and Rodriguez-Ubis J C 2004 *Chem. Mat.* **16** 1517-1522; Brunet E, Alhendawi, H M H, Juanes O, Jimenez L and Rodriguez-Ubis J C 2009 *J. Mat. Chem.* **19** 2494-2502.

[15] Hoertz P G and Mallouk T E 2005 *Inorg. Chem.* **44** 6828-6840.

[16] Brunet E, Alonso M, Mata M J, Fernandez S, Juanes O, Chavanes O and Rodriguez-Ubis J C 2003 *Chem. Mat.* **15** 1232-1234; Brunet E, Alonso M, Cerro C, Juanes O, Rodriguez-Ubis J C and Kaifer A E 2007 *Adv. Funct. Mat.* **17** 1603-1610.

[17] Brunet E, Alonso M, Quintana M C, Atienzar P, Juanes O, Rodriguez-Ubis J C and Garcia H 2008 *J. Phys. Chem. C* **112** 4029-4032.

[18] Teruel L, Alonso M, Quintana M C, Salvador A, Juanes O, Rodriguez-Ubis J C, Brunet E and Garcia H 2009 *Phys. Chem. Chem. Phys.* **11** 2922-2927.

[19] Brunet E, Cerro C, Juanes O, Rodriguez-Ubis J C and Clearfield A 2008 *J. Mat. Sci.* **43** 1155-1158.

[20] Brunet E, Alhendawi H M H, Cerro C, Mata M J, Juanes O and Rodriguez-Ubis J C 2006 *Angew. Chem., Int. Ed.* **45** 6918-6920.

[21] Salonen J, Kaukonen A M, Hirvonen J and Lehto V-P 2008 *J. Pharm. Sci.* **97** 632-53.

Magnetic Properties of Glass Ceramic in Fe_3O_4-MnO_2-SiO_2 System

Takayuki Abe[1], Tetsuo Kisi[1,2], and Atsuo Yasumori[1,2]

[1] Department of Materials Science and Technology, Tokyo University of Science, 2641 Yamazaki, Noda, 278-8510 Chiba, Japan

[2] Polyscale Technology Research Center, Tokyo University of Science 2641 Yamazaki, Noda, 278-8510 Chiba, Japan

yasumori@rs.noda.tus.ac.jp

Abstract. The magnetic glass-ceramics was prepared by a melt-quenching method with the phase-separated melts in two-liquids stable immiscibility region of Fe_3O_4-MnO_2-SiO_2 system. The obtained glass-ceramics had spinel single phase of Fe_3O_4-MnO_2 solid solution, and showed ferromagnetic properties. The saturated magnetization became a maximum by adding a small amount of MnO_2 with low coercive force. These properties of both high magnetization and low coercive force by Mn doping are very effective to realize the magnetic materials for the hyperthermia treatment.

1. Introduction

The application of magnetic materials has been widely spreading, nowadays. As we have already mentioned in the previous work [1], hyperthermia treatments of tumors using ferro-, ferri- or superparamagnetic plate, needle or particles are one of the treatments for cancer [2-8]. As the basic study of the magnetic materials for such an objective, we have investigated the phase-separated glass-ceramics in Fe_3O_4-TiO_2-SiO_2 system prepared by a melt-quenching method because the melt-quenching process is feasible to control the shape of material such as a wire and a droplet and the phase separation texture has advantageous to protect the magnetic fine crystals against chemical reaction by a silica-rich glass phase. The obtained glass-ceramics showed ferromagnetic and electric conductive properties [1].

In magnetic oxide materials, a magnetic hysteresis loss is dominant origin of a generation of heat for the hyperthermia treatment. Thus, both a large hysteresis loop of a ferromagnetic material and a high frequency of an applied magnetic field are desirable characters and conditions to generate a large amount of heat. However, a high applied magnetic field with a high-frequency may give adverse effects to patient's health. Thus, the magnetic materials should have large saturated magnetization and appropriately low coercive force to realize an enough heat generation for the hyperthermia without trouble to a patient's health.

On the other hand, the manganese doped ferrite is reported to show a relative high magnetization among various ferrites [9, 10]. Furthermore, both Fe_3O_4-SiO_2 and MnO_2-SiO_2 systems have stable immiscibility region above their liquidus temperatures [11, 12]. Therefore, in this study, we prepared the magnetic glass-ceramics in the Fe_3O_4-MnO_2-SiO_2 system by use of a melt-quenching technique. The precipitated crystalline phase, the internal texture and the magnetic property of the melt-quenched samples were examined.

2. Experimentals

In the compositions of $(40\text{-}x)Fe_3O_4\text{-}xMnO_2\text{-}60SiO_2\text{+}5Al_2O_3$ (mol%, x=0, 8, 16, 24), the starting raw materials of $\alpha\text{-}Fe_2O_3$, MnO_2, SiO_2 (α-quartz), Al_2O_3 (corundum) were mixed by ball-milling for 24 hours with ethanol. After drying the mixture of raw materials, the obtained powders were molded into a rod shape by use of a cold isostatic press, and the rods were sintered at 1200 °C for 12 hours in air. The sintered samples were melted for 3 minutes in air by use of an infrared imaging furnace (ULBAC, PRC MR-H500) and subsequently quenched.

The crystalline phases in the melt-quenched samples were analyzed by a powder X-ray diffraction (XRD, SHIMADZU, XRD-6100). The magnetic properties of the samples were evaluated from their magnetization curves, which were measured by use of a vibrating sample magnetometer (VSM, Riken Denshi Model BHV-55). The melt-quenched samples were cut vertically and polished, and their phase-separation textures were observed by a scanning electron microscope (SEM, JEOL, JSM-5800LV).

3. Result and discussion

3.1. Crystalline phase

The XRD patterns of the melt-quenched samples are shown in Figure 1. The sharp diffraction peaks of each sample were identified as those from spinel phase of Fe_3O_4 and/or $Fe_3O_4\text{-}MnO_2$ solid solution. In order to examine the effect of Mn doping on the crystal structure of ferrite, the changes of *d*-values of (220) and (311) diffraction peaks and the changes of integral intensities of those peaks against Si (111) as an internal standard substance were estimated and those results are shown in Figure 2. The former results indicate that both diffraction peaks linearly shifted to low angle with increase of the amount of MnO_2. This result means that Vegard's rule is applicable to $Fe_3O_4\text{-}MnO_2$ solid solution in this system, which was caused by replacing Mn ion with Fe ion. On the other hand, the later ones showed that the amount of precipitated spinel phase slightly decreased with increase of the amount of MnO_2. This may indicate that the Fe atoms tend to form not spinel structure but glassy structure with SiO_2 component as the amount of MnO_2 increased.

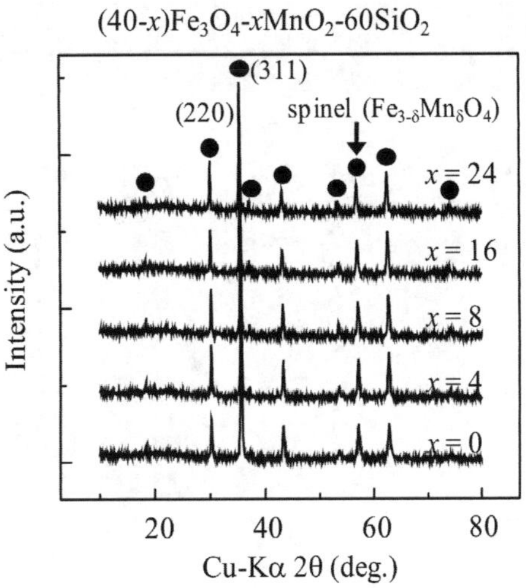

Figure 1. XRD patterns of the melt-quenched samples.

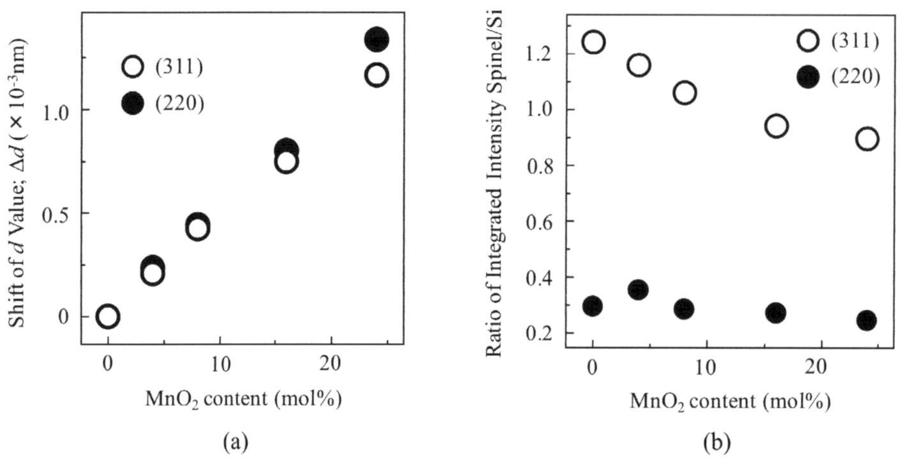

Figure 2. Changes of the *d* values; (a) and the diffraction intensities; (b) of precipitated spinel phase.

3.2. Internal texture

The SEM photographs of the melt-quenched samples are shown in Figure 3. The bright and dark areas are attributed to Fe-Mn-rich and SiO_2-rich phases, respectively. At the near-surface part of the melt-quenched droplet, the phase-separated texture was clearly observed and it became fine with increase of MnO_2 content, however, the phase separation texture was not observed inside of the melt-quenched droplets. On the other hand, the Fe-Mn rich phase was well-crystallized, especially in the samples of x=0 and 8 mol%. These results correspond to the result of the amount of crystalline phase in the XRD measurement shown in Figure 2 (b).

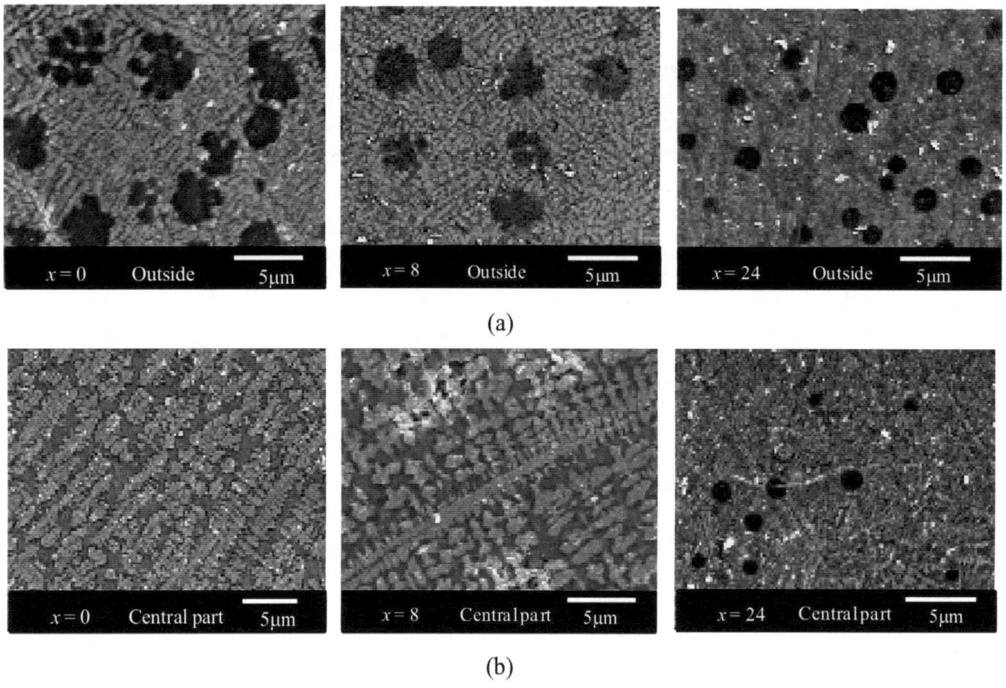

Figure 3. Cross-sections of the melt-quenched samples: the near-surface parts; (a) and the inside; (b).

3.3. Magnetic properties

The magnetization curves of the melt-quenched samples are shown in Figure 4. Figure 5 (a) shows the change of the magnetization at maximum applied magnetic field of 1.193×10^6A/m and the amount of precipitated spinel phase. Figure 5 (b) shows the coercive force (H_C) and the crystallite diameter estimated from the diffraction peak (311) and Scherrer's equation. As the amount of the precipitated spinel phase increased, the magnetization also increased. However, the sample of x=4 mol% exhibited the highest magnetization (saturated and residual). This result indicates that the magnetization increased by the small doping of Mn ions into spinel structure but decreased further doping of Mn ions. On the other hand, the H_C and the crystallite size of each sample had some relationship, that is, a large crystallite size induced a low coercive force, which was caused by the formation of multiple magnetic domains in the large crystal. The obtained H_C in this system is much lower than that of Fe_3O_4-TiO_2-SiO_2 system ($30Fe_3O_4$-$10TiO_2$-$60SiO_2$ mol%, 3.47×10^4A/m), though the crystallite diameter; 36.5 nm is similar to those of the present samples.

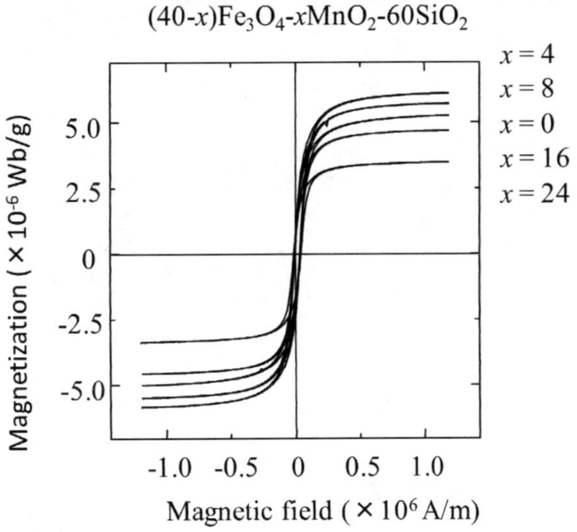

Figure 4. Magnetization curves of the melt-quenched samples.

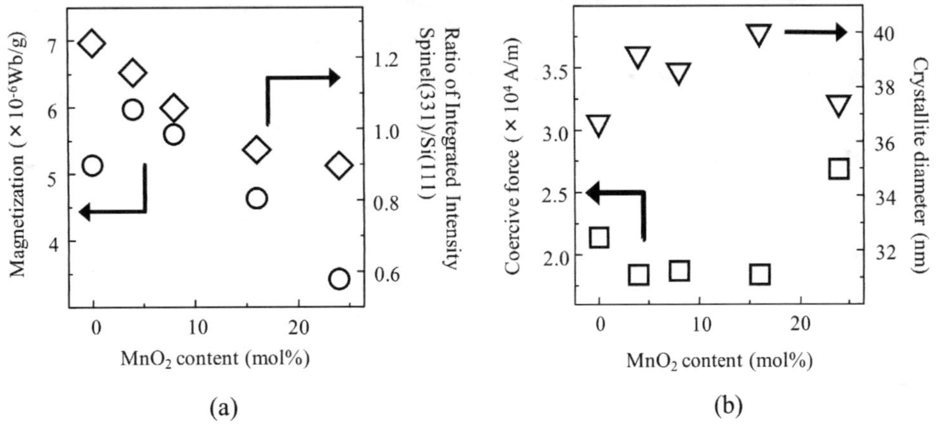

Figure 5. Change of the magnetization and the coercive force with MnO_2 content.

4. Summary

The magnetic glass-ceramics was prepared by quenching the two-immiscible phase-separated melt in the Fe_3O_4-MnO_2-SiO_2 system. The precipitated crystalline phase was spinel of Fe_3O_4 and/or the Fe_3O_4-MnO_2 solid solution. The formation of solid solution, in which Mn ions doped into spinel phase, was confirmed by the shift of the diffraction peak. In the sample of lower MnO_2 content, the larger amount of spinel was precipitated. The phase-separated texture was observed in all samples, and the crystallized texture was observed in the samples of lower MnO_2 content, especially. The saturated magnetization once increased and showed maximum at x=4 mol% and subsequently decreased with increase of the amount of MnO_2 content. The coercive force became smaller than that of Fe_3O_4-TiO_2-SiO_2 system. Though the coercive force did not change largely by changing the MnO_2 content, there was some relationship between coercive force and the spinel crystallite size, which caused by the formation of the multiple magnetic domain structure. These results show that the Mn doping in this system is very effective to realize the magnetic materials having both a high magnetization and a low coercive force, and that the obtained material will be possible to suppress the burden of patients in the hyperthermia treatment because of using lower applied magnetic field.

Acknowledgements

This work was supported by "Academic Frontier" project for private universities: matching fund subsidy from MEXT (Ministry of Education, Culture, Sports, Science and Technology), 2006-2010.

References

[1] Kono K, Kishi T, Ohgaki T and Yasumori A 2008 *J. Phys: Conference Series* **106** 012010.

[2] Atsumi T, Jeyadevan B, Sato Y and Tohji K 2006 *J. Magn. Soc. Jpn.* **30** 555-560.

[3] Kong G, Braun R D and Dewhirst M W 2000 *Cancer Research* **60** 4440-4445.

[4] Ohura K, Ikenaga M, Nakamura T, Yamamuro T, Ebisawa Y, Kokubo T, Kotoura Y and Oka M 1991 *J. Appl. Biomater.* **2** 153-159.

[5] Xiang S –H and Liu J 2008 *Int. J. Heat & Mass Transfer* **51** 5486-5496.

[6] Arcos B, Real R P del and Vallet-Regi M 2002 *Biomaterials* **23** 2151-2158.

[7] Shimizu T, Asano H, and Matsui M 2006 *J. Magn. Soc.Jpn.* **30** 166.

[8] Bahadur D and Giri J 2003 *Sadhana* **28** 639-656.

[9] Hiraga T, Okutani K and Ojima T 1986 *Ferrite* (Maruzen) pp 9-35.

[10] Eba H and Sakurai K 2007 *Chem. Anal., Japan* **38** 109-119.

[11] Phillips B and Muan A 1959 *J. Am. Ceram. Soc.* **42** 431-423.

[12] White J, Howat D D and Hay R 1934 *J. R. Tech. Coll. (Glasgow)* **3** 231-240.

Protein patterning by atmospheric-pressure plasmas

A Ando[1], M A Sayed[2], T Asano[3], R Tero[3], K Kitano[1], T Urisu[3] and S Hamaguchi[1]
[1] Center for Atomic and Molecular Technologies, Osaka University, Osaka, 565-0871, Japan
[2] Department of Structural Molecular Science, The Graduate University for Advanced Studies, Aichi, 444-8585, Japan
[3]Division of Biomolecular Sensing, Department of Life and Coordination-Complex Molecular Science, Institute for Molecular Science, Aichi, 444-8585, Japan
E-mail : ando@ppl.eng.osaka-u.ac.jp

Abstract. Low-frequency (LF) atmospheric-pressure plasma (APP) jets have been used for patterning of thin fibronectin films deposited on a silicon (Si) wafer with the use of a metal stencil mask. Since fibronectin is an adhesion protein that can be found in the extracellular matrix (ECM), a micro-patterned fibronectin film may be used for arranging living cells in a desired pattern on a surface. Removal of fibronectin from the surface by plasma application was observed by Fourier transform infrared spectroscopy (FT-IR) and atomic force microscopy (AFM). It has been found that removal of fibronectin takes place even in the location away from the direct plasma jet application spot, which indicates that desorption of fibronectin is likely to be caused by chemically reactive charge-neutral species that can diffuse away from the plasma without emitting visible light.

1. Introduction

Biosensor and biochip technologies have been widely used for functional analyses of biological materials and systems such as DNAs, proteins and cells. The use of biosensors and biochip is expected to expand further as the technologies are further advanced [1-3]. Some biosensors and biochips require micro-patterning technologies to arrange and immobilize biological materials and living cells on a substrate surface while maintaining their biological functions. Especially of interest recently have been micro patterning techniques for large areas as the sizes of biochips continue to increase.

Part of a biochip that serves to transmit and receive electrical signals may be more easily fabricated on a silicon (Si) substrate, to which existing technologies for semiconductor manufacturing can be applied directly for fabrication of such systems. On the other hand, few biological materials or living cells can be immobilized directly on a Si surface. For example, most living cells can be immobilized on a Si substrate only if an extracellular matrix (ECM) layer is placed on the Si substrate. The main functions of ECMs are to provide support and anchorage to cells and to regulate intercellular communication. An ECM is typically composed of proteins such as collagen, laminin, fibronectin, *etc.*

Micro-patterning of ECMs has been mostly performed so far by micro contact printing (μCP) with the use of polydimethylsiloxane (PDMS) stamps [4, 5] or ink-jet printing [6] with the use of an ink-jet printer. These techniques are widely employed in the biochip research community. With the μCP technique, for example, one can print protein in the grid patterns of several square micrometres on a several square millimetre substrate. However, for some applications, efficient patterning on a larger substrate (e.g., with the area of several square centimetres) is desirable. As we shall see in this article, protein pattering by atmospheric pressure plasmas (APPs) may provide a solution to such demand.

In the present work, therefore we have examined a possibility of developing a method for ECM patterning on an area of several square centimetres, using low-temperature APP. Such plasmas can be chemically highly reactive and impose little thermal load on the substrate. The method may remind one of the plasma etching technologies widely used in the semiconductor industry. However, the method explored here can be used in air without requiring expensive vacuum equipment. As a mask for patterning, we have used a stainless-steel stencil, a far less expensive alternative to photoresist. Protein patterns and surface characteristics of the substrate after plasma exposure were analyzed by cell culture, Fourier transform infrared spectroscopy (FT-IR) and atomic force microscopy (AFM).

2. Experimental

2.1. Low-frequency atmospheric-pressure plasma jets

Low-frequency (LF) APP jets are discharges generated in air when a pulsed high voltage is applied to a helium (He) gas flow that flows inside a glass tube, as shown in figure 1. The system is essentially the same as those given in Refs. [7, 8]. In our experiments, the peak-to-peak voltage applied to the electrode wound around the glass tube was approximately 7 kV and the pulse frequency was10 kHz. The flow rate of He gas was 5 L/min. The inner diameter of the glass tube used for protein patterning was 4 mm and that for surface analysis was 8 mm. The distance between the sample substrate and the end of glass tube was set at 20 mm.

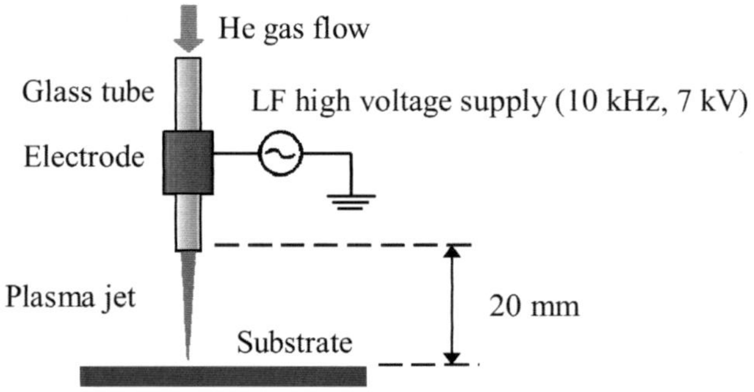

Figure 1. Low frequency atmospheric pressure plasma (LF-APP) jets.

2.2. *Sample*

In this work, a fibronectin solution (200 μg/ml) was used to form a fibronectin coating of a substrate. Fibronectin, which is found in ECM and body fluids, is a multifunctional glycoprotein composed of subunits with a molecular weight of approx 250 kilodaltons (kDa) [9]. Fibronectin is often used as an ECM when cells are cultivated on SiO_2 substrates [10]. After some quantity of the fibronectin solution was dropped onto the substrate to cover a sufficient area of its surface and let adsorbed for 1.5 hour at room temperature, extra fibronectin was rinsed off by ion-exchanged water. The substrate used in our experiments was either (naturally oxidized) Si for patterning tests and AFM observation or gold for FT-IR analysis.

3. Experimental results

3.1. Protein patterning by LF plasma jets

LF plasma jets were applied to a 15×15 mm square Si substrate coated with fibronectin on which a stainless-steel stencil mask was directly placed. The mask has three different sets of slits, as shown in figure 2; (1) ten 100 μm-wide slits with the distance between two successive slits of 100 μm, (2) ten 100 μm-wide slits with the distance between two successive slits of 200 μm and (3) ten 200 μm-wide slits with the distance between two successive slits of 100 μm. In the experiments shown in this section, LF jets were applied only to the area (1) (i.e., ten 100 μm-wide 100 μm-apart slits) for 20 sec. After the plasma irradiation, HEK293 cells (Human Embryo Kidney cells) were cultivated on the entire sample surface. Adhesion, division and alignment of cells on the substrate were observed at every 24 hours after cell culture.

Shown in figure 3 is a photograph of the sample substrate surface (without the stencil mask) taken 72 hours after cell culture. Cells are confluently cultured on the Si substrate and appear white in colour by the naked eye. Although it is not very clear in figure 3, the area (1) appears darker, which indicates the Si substrate surface is seen and few cells were grown in this area. On the other hand, the white lines observed in the area (3) are cells aligned along the slit patterns. A micrograph of each area is given in figure 4, which shows that few cells indeed exist in the area (1) and cells are only sparsely distributed in the area (2) whereas cells align along the mask slits in the area (3).

Since cells are expected to grow only on the fibronectin coated area, the results shown here indicate that the duration of plasma application was too long, so that fibronectin in the pattern area (1) was completely eliminated even under the mask whereas charge-neutral chemically reactive species (such as gas-phase free radicals) that diffused away from the plasma jet (and therefore were sufficiently diluted in air) patterned the fibronectin film along the mask slits in the area (3).

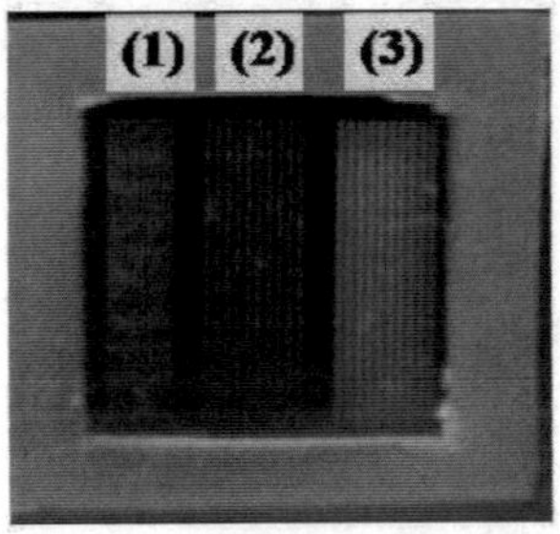

Figure 2. The stencil mask with slits used in this work.

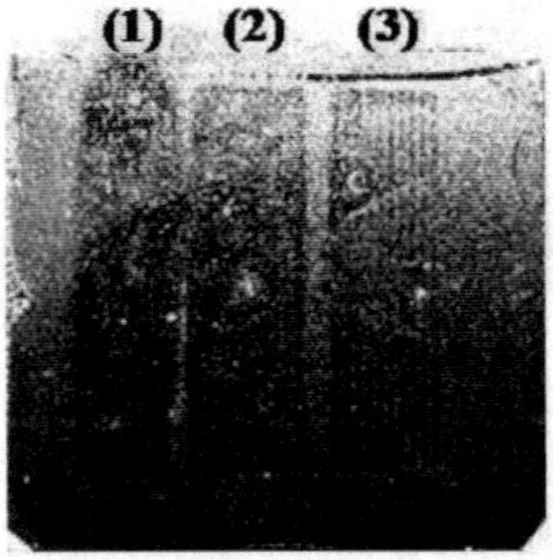

Figure 3. A substrate surface photograph taken 72 hours after cell culture. LF plasma jets were applied only to the area (1). Cells appear white by the naked eye.

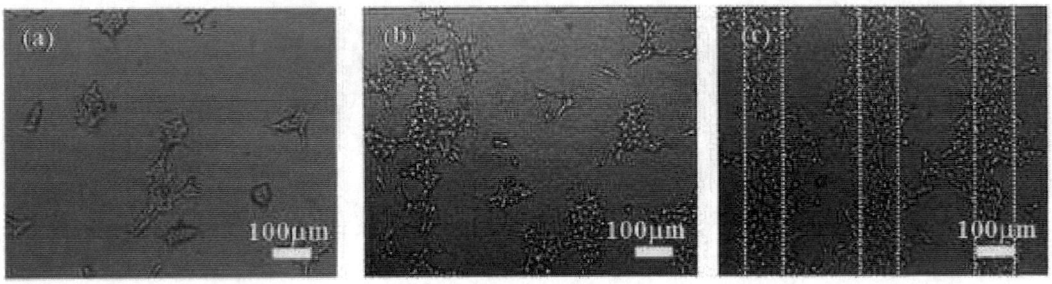

Figure 4. Micrographs of the cell arrangement patterns on the Si substrate of figure 3. The micrograph (a) is in the area (1), (b) in the area (2) and (c) in the area (3). The dashed lines of (c) indicate slit patterns. Cells are expected to grow only on fibronectin-coated areas.

3.2. Surface analysis by Fourier transform infrared spectroscopy (FT-IR)
Fibronectin was deposited on a 13×13 mm square gold substrate which was formed by coating Cr (a few nanometres thick) and Au (45 nm thick) on a glass substrate for FT-IR analysis [11]. LF plasma jets applied to the substrate was scanned uniformly over the entire substrate surface. No stencil mask was placed on the substrate. The plasma application time was varied from 0 sec to 20 sec.

The amid I band spectra (1600 ~ 1700 cm^{-1}) of the fibronectin coating obtained from FT-IR are shown in figure 5 as a function of the plasma application time. The amid I band spectra indicate the presence of the secondary structures such as α-helix, β-turn and β-sheet of protein [11-13]. Also shown by a dotted curve in figure 5 for comparison is the amid I band spectra of the fibronectin coated substrate to which only a He gas flow (without a plasma) was applied for 20 sec. It is shown that, as the plasma-application time increases, the amid I band spectra diminish and nearly vanish at around 20 sec of plasma application. It should be noted that the spectral shape of the amid I band hardly changes as it diminishes, which may indicate disappearance of fibronectin from the surface rather than decomposition of its secondary structures.

3.3. Surface analysis by atomic force microscopy (AFM)
Figure 6 shows the sample substrate surface morphologies observed by AFM (a) before and (b) after plasma application of 120 sec. The entire surface of fibronectin-coated Si substrate was uniformly irradiated by the scanning plasma jets. The average roughness is 3.0 nm for (a) and 0.65 nm for (b). The latter is nearly equal to the roughness of an uncoated Si substrate (which is typically 0.5 nm). From this result together with the FT-IR observation discussed in the previous subsection, it is highly likely that fibronectin is completely desorpted by the plasma application of a sufficient duration time.

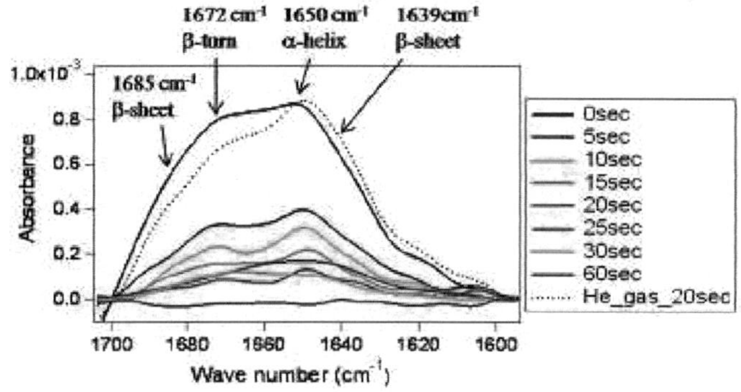

Figure 5. Spectra of the amid I band (1600 ~ 1700 cm^{-1}) measured by FT-IR.

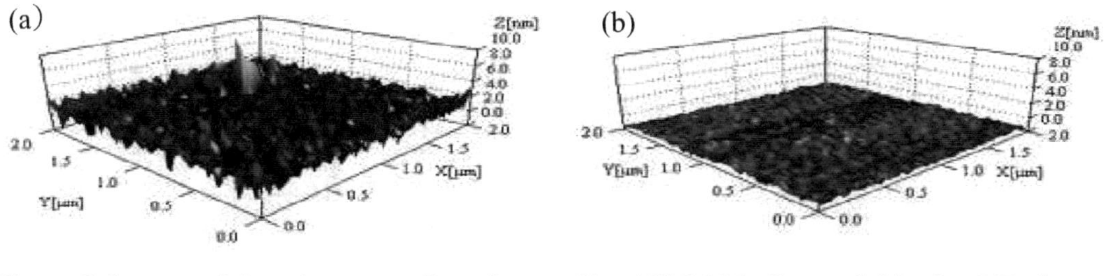

Figure 6. Images of the substrate surface observed by AFM (a) before and (b) after LF-plasma jet application for 120 sec.

4. Discussion
Fibronectin films coated on Si substrates have been patterned by LF plasma jets in this study. Our experimental observations based on cell culture, FT-IR, and AFM have indicated that chemically

reactive gas-phase species (such as free radicals) generated by the plasma are likely to react with fibronectin and remove it from the surface. The reactive species, however, have not been identified in this study.

In the process presented in this work, surface patterns of a fibronectin film are provided by a stainless-steel stencil mask. Since a thin stainless-steel sheet can be easily patterned down to a few dozen micrometres in size by mechanical means, the mask used in our work is inexpensive and reusable. With the use of such a mask, fibronectin directly exposed to the chemically reactive species generated by the plasma is readily removed from the surface whereas fibronectin covered by the mask is less easily removed.

Since the stencil mask is simply placed on the fibronectin film mechanically (with the periphery of the mask pinned down to the substrate by glue), there must be some small spatial gap between fibronectin and the stencil, into which the gas-phase reactive species can diffuse. Therefore, so-called "under-etch", i.e., removal of fibronectin under the stencil mask was also observed in our study when the plasma exposure time was sufficiently long. The fact that no cell grew in the area (1) of figure 3 is the indication of such "under-etch," where it seems fibronectin on the substrate surface was completely removed due to overexposure to the plasma despite the presence of a patterned stencil mask.

5. Conclusions

LF plasma jets have been used to pattern fibronectin films deposited on Si substrates. The present study has clearly indicated that a process employing a low-temperature APP with a metal stencil mask is a feasible method to pattern a large area of ECM protein films at least up to several square centimetres. Optimization of plasma application conditions (such as the distance between the plasma jet and the substrate, plasma exposure time, supplied gas specie, and gas flow rate) and mechanism analysis of patterning (influencing radicals and chemical reaction) are deferred to future studies.

Acknowledgments

This work was supported by the Joint Studies Program (2009 ~ 2010) of the Institute for Molecular Science.

References

[1] Yamato M, Konno C, Utsumi M, Kikuchi A and Okano T 2002 *Biomaterials* **23** 561

[2] Jiang X Y, Ferrigno R, Mrksich M and Whitesides G M 2003 *J. Am. Chem. Soc.* **125** 2366

[3] Zhang Z L, Asano T, Uno H, Tero R, Suzui M, Nakao S, Kaito T, Shibasaki K, Tominaga M, Utsumi Y, Gao Y L, and Urisu T 2008 *Thin Solid Films.* **516** 2813-2815

[4] Bernard A, Delamarche E, Schmid H, Michel B, Bosshard H R and Biebuyck H 1998 *Langmuir* **14** 2225

[5] Nishizawa M, Kaji H, Takoh K and Matsue T 2004 *Journal of Surface Science Society of Japan* **25**(5) 290

[6] Morozov V N and Morozova T Y 1999 *Analytical Chemistry* **71** 3310

[7] Kitano K and Hamaguchi S 2008 *The Japan Society of Applied Physics* **77**(4) 383

[8] Ikawa S, Kitano K, and Hamaguchi S 2010 *Plasma Proc. Polym.* **7**(1) 33

[9] Akiyama S K, Nagata K and Yamada K M 1990 *Biochim. Biophys. Acta* **1031** 91

[10] Urisu T, Asano T, Zhang Z, Uno H, Tero R, Junkyu H, Isoda H, Arima Y, Iwata H, Shibasaki K and Tominaga M 2008 *Anal. Bioanal. Chem.* **391** 2703

[11] Sayed A M, Uno H, Harada K, Tanaka K, Kim Y H, Nakaoki Y, Okumura K, Tero R and Urisu T 2008 *Chem. Phys. Lett.* **46** 235

[12] Koteliansky V E, Glukhova M A, Bejanian M V, Smirnov V N, Filimonov V V, Zafite O M and Venyaminov S Y 1981 *Eur. J. Biochem.* **119** 619

[13] Venyaminov S Y, Metsis M L, Chernousov M A and Koteliansky V E 1983 *Eur. J. Biochem.* **135** 485

Effect of light irradiation from inductively coupled Ar plasma on etching yields of SiO_2 film by CF_3 ion beam injections

S Yoshimura[1], Y Tsukazaki[1], K Ikuse[1], M Kiuchi[1,2] and S Hamaguchi[1]

[1]Center for Atomic and Molecular Technologies, Graduate School of Engineering, Osaka University, 2-1 Yamadaoka, Suita, Osaka 565-0871, Japan

[2]National Institute of Advanced Industrial Science and Technology (AIST), 1-8-31 Midorigaoka, Ikeda, Osaka 563-8577, Japan

E-mail: yosimura@ppl.eng.osaka-u.ac.jp

Abstract. Effects of light irradiation on etching yields of SiO_2 films by CF_3 ion beams have been investigated with a low-energy mass selected ion beam system. An inductively coupled plasma (ICP) from Ar gas was used as the light source. The spectrum of light emitted from the Ar ICP was measured by vacuum ultraviolet and visible monochromators. Etching yields of SiO_2 films by 470 eV CF_3 ion beam injections were obtained from the ratio of the number of incident CF_3 ions to that of removed SiO_2 atoms during the ion beam injection process with or without simultaneous light irradiation from the Ar ICP. The incident ion number was evaluated from the integration of temporal evolution of ion beam current. The number of removed atoms can be obtained from the etched depth of SiO_2 films measured by a laser interferometer. It is found that the obtained SiO_2 etching yield by simultaneous irradiation of CF_3 ions and photons from the light source was smaller than that by ion beam irradiation only.

1. Introduction

Carbon tetrafluoride (CF_4) gases are widely used in semiconductor etching processes. Etching yields of various materials due to fragment ions produced in CF_4 gas-based plasmas provide useful information on the control of manufacturing processes. The fragment elements produced from CF_4 gas and their kinetic energies play an important role in the etching processes. The etching yields of silicon dioxide (SiO_2) films by fragment ions (F^+, CF^+, CF_2^+, and CF_3^+) from CF_4 gas in the energy range of 250-2000 eV have been measured with ion beam devices [1-3].

In addition to the ion energy, ultraviolet (UV) light, which can cause excitation of substrate atoms and therefore change their average bond energies, may play an important role in CF_4 gas-based plasma processes [4, 5]. Although it is important to evaluate effects of UV light on etching processes, the influence of Ar inductively coupled plasma (ICP) light has not been extensively investigated yet so far. In this experiment, SiO_2 films were irradiated by light from an Ar ICP in addition to a CF_3 ion beam for the investigation of the influence of UV light on the etching yields.

2. Experimental apparatus

This experiment was carried out in a low-energy mass selected ion beam system [6]. This system consists of a Freeman-type ion source, an extractor electrode, a magnetic-field-based mass selector, a

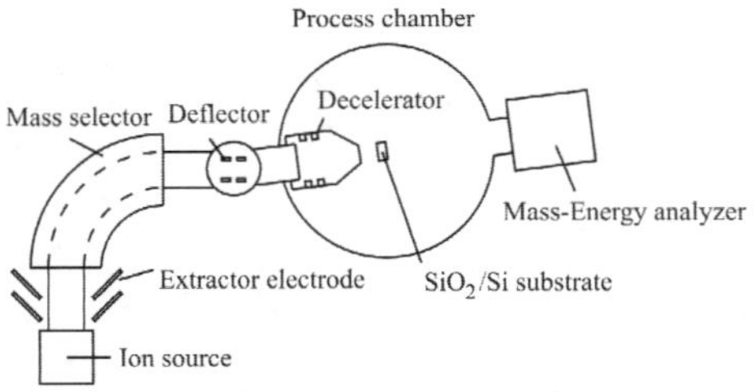

Figure 1. A schematic drawing of the low-energy ion beam system.

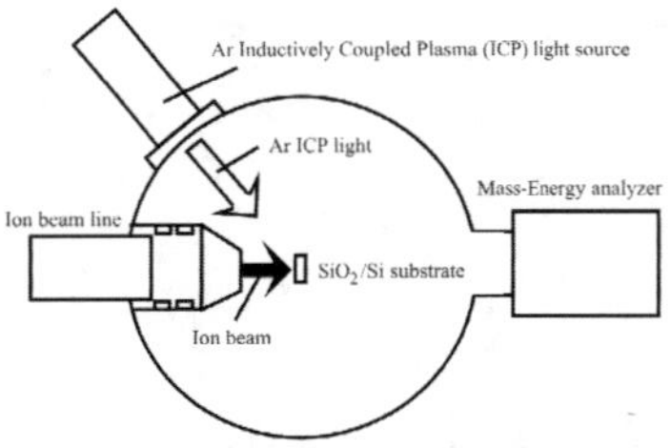

Figure 2. A schematic diagram of the process chamber of the low-energy ion beam system.

deflector, a decelerator, a process chamber, and a mass-energy analyzer, as shown in figure 1. From the source gas (Ar-diluted CF_4 gas), F, CF, CF_2, CF_3 ions were produced in addition to Ar ions in the ion source chamber. These ions were extracted by a high voltage of -25 kV applied to the extractor electrode. The ions were separated by the mass selector, so that only desired ion species (CF_3 ions in the present experiment) remained in the beam line without impurity ions. The mass-selected ion beam was then deflected by a small angle, the process of which removed high energy neutral species generated by charge exchange reactions with residual gases in the beam line. Finally, the ion beam was decelerated to the desired kinetic energy and injected to the SiO_2 thin film formed thermally on a Si substrate. The ion injection angle was normal to the substrate. The quality of the ion beam (i.e., energy and mass distribution of incident ions) was examined by the mass-energy analyzer (balzers, PPM-421).

The substrate was set at the sample holder on the manipulator. The sample holder was connected with an ammeter (KEITHLEY, 6485). A Mo orifice plate which was connected with the ground was set at 4 mm in front of the substrate surface. The diameter of the orifice was 7.5 mm. The ion beam irradiated the substrate surface through the orifice, the current of which was measured by the ammeter. The substrate was set at room temperature.

Etching yields of the SiO_2 film were determined from the ratio of the number of incident CF_3 ions to that of removed SiO_2 atoms during the ion beam injection process. The incident ion number was evaluated from temporal evolution of the ion beam current. The number of removed atoms can be obtained from the etched depth of SiO_2 film measured by a laser interferometer.

In the present experiment, the etching yield due to ion injections only was compared with that due to simultaneous ion and light irradiation. The substrate surface was irradiated with a light source at an oblique angle (50 degree), as shown in figure 2. In the present experiment, the light source was an Ar ICP. The base pressure in the Pyrex chamber (50 mm in diameter, 350 mm in length) of ICP device was about 0.4 Pa and the pressure of Ar gas was 4-5 Pa. The Ar plasma was produced by a 13.56 MHz power source (NIHON KOSHUHA, HFS-005) at 16-30 W. The distance between the ICP device and the substrate was about 40 cm. The Ar ICP light was set to pass through an MgF_2 window, which allows light with the wavelength of 110 nm or longer to pass through.

3. Experimental results

Firstly, the spectrum of light emitted from the Ar ICP was measured. The Ar ICP light had a wide spectrum including UV and visible light. Figure 3 shows a vacuum ultraviolet emission spectrum (110-290 nm) of the Ar ICP measured by a Seya-Namioka monochromator which was connected with the ICP device through the MgF_2 window. Line spectra of impurities (hydrogen H, carbon C, nitrogen N, and oxygen O) are seen in the 110-200 nm region while the emission is dominated by NO(γ) bands in the 200-290 nm region [7]. It is seen that there are no dominant line spectra of Ar in this region. The UV light intensity measured by a UV power meter (HAMAMATSU, H8025-146, sensitive range = 120-200 nm) were found to be 8 $\mu W/cm^2$. Figure 4 shows an emission spectrum of the Ar ICP

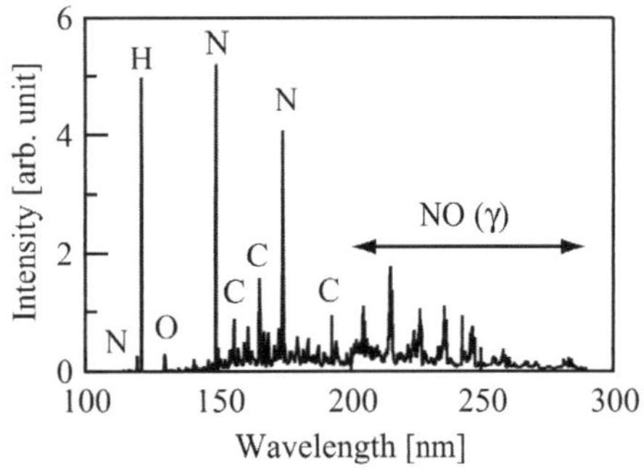

Figure 3. The vacuum ultraviolet emission spectrum of an inductively coupled plasma produced from Ar gas.

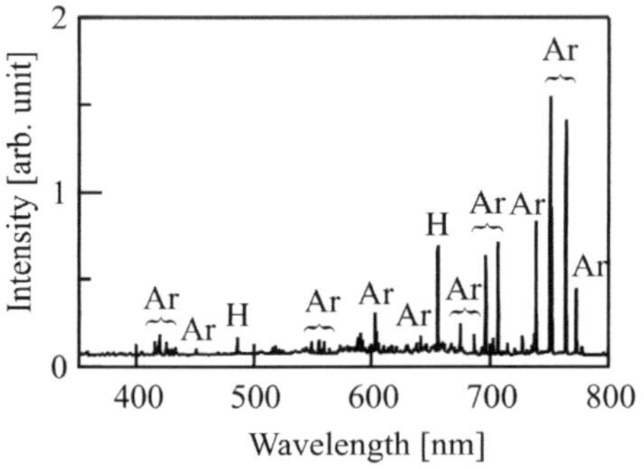

Figure 4. The visible light emission spectrum of an inductively coupled plasma produced from Ar gas.

measured through the Pyrex chamber by a monochromator (Ocean Optics, HR4000) in the visible region (350-800 nm). Line spectra of Ar and H are seen in figure 4.

A mass spectrum in the ion source was measured by scanning the mass selector current. It was found that the intensity of CF_3 ions was the largest among all fragment ions produced from CF_4 gas. Therefore, we select CF_3 as the injection ions.

The energy distribution of CF_3 ions was measured by PPM-421. In this case the peak energy of the CF_3 ion was found to be 470 eV. The full width at the half maximum (FWHM) of the distribution was about 5 eV.

Figure 5 shows the temporal evolution of the ion beam current during the etching process of a SiO_2 film by CF_3 ions with Ar ICP light. The beam current of CF_3 ions was 0.18 μA and almost constant during the process. The power of ion beam was 190 $\mu W/cm^2$, which was larger than the power of UV light (8 $\mu W/cm^2$). The injecting particle number of CF_3 ions was evaluated by integrating temporal evolution of the beam current.

After the ion beam injection experiment, the etched depth of each SiO_2 film was measured by a laser interferometer. The thickness of SiO_2 film before the ion beam injection was 103 nm. Figure 6 shows the profile of a SiO_2 film thickness after ion beam injections. A hollow region due to the etching of SiO_2 film can be seen near the center in figure 6. The diameter of etched region is about 8 mm, corresponding to the size of Mo orifice. The number of etched atoms can be calculated from the volume integration of etched depth and the density of SiO_2 film.

Etching yields of the SiO_2 film were calculated from the ratio of the number of injected CF_3 ions to that of etched SiO_2 atoms. The etching yield of SiO_2 by the CF_3 ion beam *without* Ar ICP light was evaluated to be 1.1. On The other hand, the etching yield of SiO_2 by the CF_3 ion beam *with* Ar ICP light was 0.85. Therefore, the etching yield of the SiO_2 film by CF_3 ions with Ar ICP light was slightly smaller than that without Ar ICP light. The reduction of etching yield by the light irradiation suggests the interaction of ion beam and Ar ICP light. When the substrate was irradiated by Ar ICP light only, etching of the film was not observed.

4. Summary

Effects of Ar ICP light on etching yields of SiO_2 films by CF_3 ions were investigated with a low-energy mass selected ion beam system. Prior to the ion beam injection experiment, the spectrum of light emitted from the Ar ICP was measured. Line spectra of impurities (H, C, N, and O) were seen in the 110-200 nm region while the emission was dominated by NO(γ) bands in the 200-290 nm region. On the other hand, line spectra of Ar and H were seen in the visible region (400-800 nm). The ion beam quality was investigated by PPM-421. It was found that the peak ion energy was 470 eV and FWHM of the energy distribution was about 5 eV. We have evaluated the etching yields of SiO_2 films by CF_3 ion beams with or without light irradiation. The etching yields were compared among the three

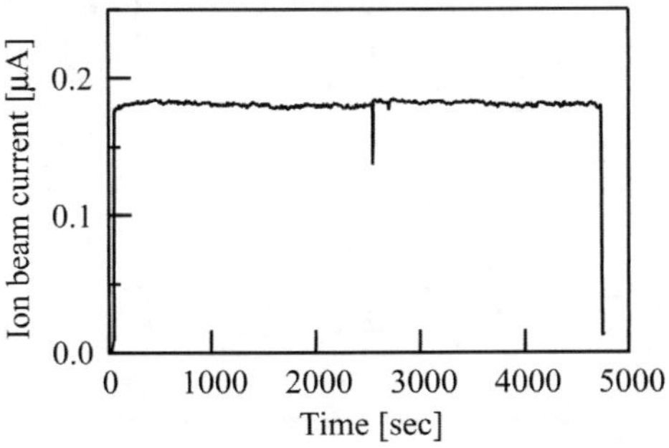

Figure 5. The temporal evolution of beam current of CF_3 ions with the irradiation of an inductively coupled Ar plasma light.

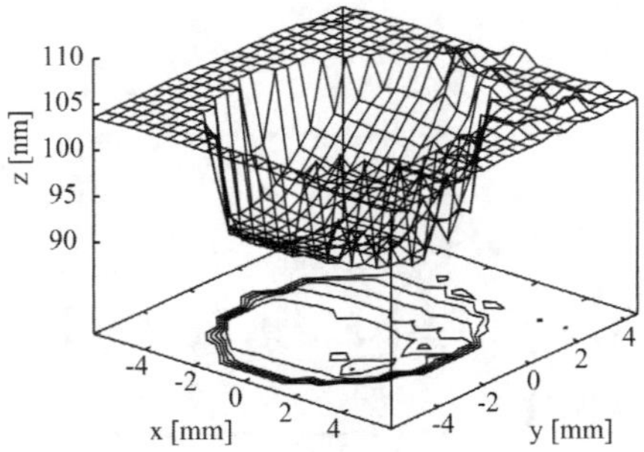

Figure 6. The thickness profile of SiO_2 film etched by CF_3 ion beam with light from inductively coupled Ar plasma.

cases: (1) ion beam injections only, (2) simultaneous irradiations with the ion beam and the light from the Ar ICP, and (3) irradiations with the light from the Ar ICP only. Prior to the experiment, we expected that the excitation of substrate materials by UV lights may result in weakening of atomic bonds temporarily and ion injections into the materials under such conditions may increase the etching yield. However, the experiment showed that the obtained SiO_2 etching yields by simultaneous irradiation of CF_3 ions and photons from the light source were smaller than those by ion beam irradiations only. This difference in etching yields may be caused by the modification of CF_x polymer formation on the substrate surface during the ion beam etching process.

Acknowledgement

This work was partially supported by a Grant-in-Aid for Scientific Research from the Ministry of Education (MEXT) of Japan.

References

[1] Ishikawa K, Karahashi K, Tsuboi H, Yanai K and Nakamura M 2003 *J. Vac. Sci. Technol.* A **21** L1

[2] Karahashi K, Yanai K, Ishikawa K, Tsuboi H, Kurihara K and Nakamura M 2004 *J. Vac. Sci. Technol.* A **22** 1166

[3] Yanai K, Karahashi K, Ishikawa K and Nakamura M 2005 *J. Appl. Phys.* **97** 053302

[4] Ishikawa K, Okigawa M, Ishikawa Y, Samukawa S and Yamasaki S 2005 *Appl. Phys. Lett.* **86** 264104

[5] Samukawa S, Jinnai B, Oda F and Morimoto Y 2007 *Jpn. J. Appl. Phys.* **46** L64

[6] Yoshimura S, Toh A, Sugimoto S, Kiuchi M and Hamaguchi S 2006 *Jpn. J. Appl. Phys.* **45** 8204

[7] Rahman A, Yalin A P, Surla V, Stan O, Hoshimiya K, Yu Z, Littlefield E and Collins G J 2004 *Plasma Sources Sci. Technol.* **13** 537

Evaluation of Si etching yields by Cl^+, Br^+, and HBr^+ ion irradiation

Tomoko Ito[1], Kazuhiro Karahashi[1], Song-Yun Kang[2], and Satoshi Hamaguchi[1]
[1]Center for Atomic and Molecular Technologies, Osaka University
[2]Tokyo Electron Ltd
E-mail : ito@ppl.eng.osaka-u.ac.jp

Abstract.
Cl_2 and HBr plasmas are widely used for poly-Si gate electrode etching processes in the semiconductor industry. In this study, sputtering yields of poly Si by Br^+ and HBr^+ ion injections, simultaneous injections of thermal H atoms and energetic Br^+ ions, and successive injections of an energetic H^+ and Br^+ ion beams, have been evaluated with the use of a multi-beam system equipped with a mass-selected ion beam injector and a H radical source. It has been shown that there is little difference in sputtering yields among all these processes, which indicates that such hydrogen has little effect on the Si sputtering yield by Br^+ ion injections. The result contrasts with an earlier speculation that hydrogen in HBr plasmas significantly enhances the Si sputtering yield by Br^+ ion injections. In addition sputtering yields of poly Si by Cl^+ ion bombardment obtained from similar beam experiments have been presented for comparison.

1. Introduction

Reactive ion etching (RIE) by HBr plasmas are widely used for poly-Si etching in semiconductor manufacturing processes. In a RIE process, various chemically reactive charge neutral species (such as atomic Br) generated in the plasma as well as ions (such as Br^+) accelerated by the sheath voltage near the substrate surface enable anisotropic etching of silicon. For the development of highly controllable etching technologies for nano-scale structures used in the modern semiconductor industry, a better understanding of interactions between Si surfaces and reactive species in plasmas is highly desirable.

Atomic hydrogen (H) is highly reactive and spontaneously reacts with Si. It has been widely believed that, in RIE processes of Si by HBr plasmas, the presence of hydrogen enhances the Si sputtering yield. Indeed the experiments by Vitale *et al.* [1] showed with the use of a plasma-beam system that the sputtering yields of poly Si by HBr plasmas is approximately twice larger than those by Br_2 plasmas under the same beam energy, which the authors of Ref. [1] attributed to the presence of hydrogen in the system.

In the plasma beam system used in Ref. [1], the beam was extracted from a plasma source and injected into the sample surfaces without mass selection. Therefore the beam contains various ionic and neutral species. Although such plasma beams may be considered to resemble the fluxes to which substrates are subject in actual plasma processing tools, it could be difficult in such systems to separate the roles that individual ionic or atomic species play for surface chemical reactions.

The goal of this study is therefore to re-examine elementary chemical reactions of HBr^+ and Br^+ ions as well as other related reactions on Si surfaces with the use of a different type of beam system. Especially of interest is to determine the role that hydrogen plays in such surface reactions. To this end, we have used a multi-beam system that can simultaneously inject a low-energy neutral beam and a mass-selected ion beam.

In this work, to examine Si etching reactions by bromine, we performed three sets of experiments. In the first set of experiments, Br^+ and HBr^+ ions were injected into Si substrates and their sputtering yields were measured as functions of the injection energy. In the second set, Si substrates were exposed to simultaneous injections of energetic Br^+ ions and charge-neutral thermal H atoms near room temperature. The flux ratio of H to Br^+ used in the experiments was about 1~10. In the third set of experiments, Si substrates were exposed to Br^+ ion injections after energetic H^+ ion injections with a sufficient fluence (dose). Injections of energetic H^+ ions are expected to damage the Si substrate, which may affect the penetration of Br^+ ions that are subsequently injected into the substrate.

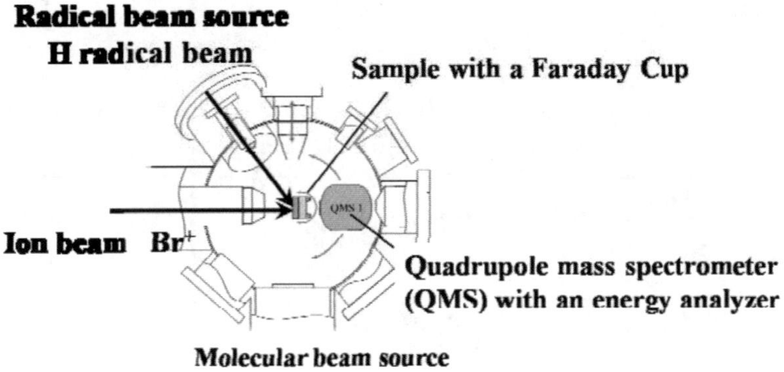

Figure 1. A schematic diagram of the reaction chamber of the multi-beam system used in this study [2].

2. Experiments

In this study, we used a multi-beam system that can irradiate a sample surface by independently controlled atoms, molecules and ions fluxes. The multi-beam system consists of a mass selected ion beam injector and a reaction chamber that is equipped with two independently controllable neutral radical/molecular beam injectors.

For the mass selected ion beam injector, a Freeman type ion source is used as the ion source, where ions are produced by arc discharges. Ions generated in the source are extracted by a high voltage and accelerated to 25keV, which sufficiently suppresses the beam emanation due to space charge. Ionic species of interest are then selected by the mass selecting magnet. The selected ions are then decelerated to a desired energy by the deceleration electrode before reaching the sample substrate in the reaction chamber. In this system, the beam energy at the substrate surface can be varied from 0 to 2keV.

Charge neutral species such as radicals that leak out of the ion source may have strong influence on etching reactions on the sample surface. In the multi-beam system, three differential pumps between the ion source and the reaction chamber reduce the density of neutral species in the beam line. The reaction chamber is always maintained in ultrahigh vacuum conditions with the pressure of about 10^{-10} Torr, which minimizes the effect of back ground gases on the sample surface. The sample substrate is set on a manipulator and can be tilted against the ion beam line, which allows a selection of the beam incident angle on the substrate. The reaction chamber is equipped with two charge-neutral radical sources; one based on a radio-frequency (RF) plasma and another for a supersonic molecular beam. These beams may be independently controlled and therefore simultaneously injected into the sample surface, as may be seen in Fig. 1. In this study, we have used the ion beam injector and the H radical source based on hydrogen RF plasmas (HD25 Oxford Applied Research). The ion injection angle was normal to the sample surface whereas the average angle of incidence for the H radical flux was 60° (measured from the normal direction of the sample). Typically the ion beam flux was 10^{13}/cm^2/sec whereas the H radical beam flux was about 10^{13}~10^{14}/cm^2/sec at the substrate surface. The kinetic energies of H radicals are typically close to that of room temperature.

The etching yield (i.e., sputtering yield) of a Si substrate is defined as the number of desorbed Si atoms per ion injection. In this study, the ion dose was calculated from the exposure time and ion current that was measured by a Faraday cup. The number of desorbed Si atoms was evaluated from the etched depth that was measured by a surface profiler (Dektak3ST). The sample used in this study was a 1.5 cm square wafer on which a 150 nm thick poly Si was formed by CVD. Before use, its native oxide was removed by hydrofluoric acid treatment. For a typical yield measurement, the ion dose on the sample surface was approximately 1×10^{17}/cm^2.

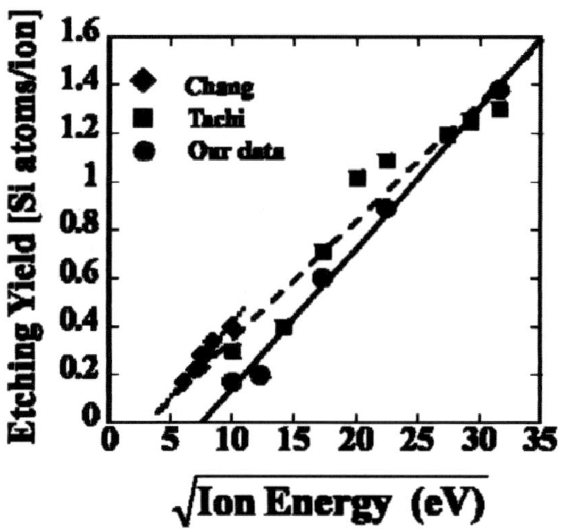

Figure 2. Sputtering yields of Si resulting from Cl^+ bombardment. The solid, broken, and dotted lines are fitting lines to the data obtained in this study (denoted by filled circles), by Chang and Sawin [3] (denoted by filled squares), and by Tachi and Okudaira [4] (denoted by filled diamonds).

3. Experimental Results

3.1 Etching by Cl^+ ion beams

Before discussing etching processes associated with bromine, we first present our measurement results of Si sputtering yields by energetic Cl^+ ions. Sputtering yields by chlorine ions and plasmas have been widely studied and the sputtering yield data that we have obtained may be compared with those obtained in the past.

Figure 2 shows sputtering yields of poly Si by Cl^+ ion injections as functions of the injection energy. The data obtained in this study are denoted by filled circles whereas data obtained by the earlier studies [3,4] are also shown for comparison. It is clearly seen that the yield obtained in this study (as well as those from the previous studies) is a linearly increasing function of the square root of the injection energy, following the well-known formula by Steinbrüchel [5].

It is also shown that, for the energy range over 300eV, the yield obtained in this study and by Tachi and Okudaira [4] are overall in good agreement. The yield data of Ref. [4] were obtained from a mass-selected ion beam system with the pressure of approximately 10^{-8} Torr during the beam injection. The pressure is higher by two orders of magnitude compared with that of this study.

The yield data by Chang and Sawin [3] shown in Fig. 2 are based on their plasma-beam experiments, where the beam was not mass selected. Their data are higher than those of [4] as well as this study. The difference may arise from the effects of neutral Cl atoms, which are known to enhance the yield significantly when they passivate the surface [3]. The reaction chamber pressure in the experiments by Chang and Sawin was about 10^{-6} Torr [3] and therefore the flux ratio of Cl atoms to Cl^+ ions is considered to have been about 0.1-0.01 on the sample surface. Since the flux ratio of Cl^+ ions to Cl atoms is 10^{-3} or less in our experiments due to the high vacuum condition of 10^{-10}Torr, the difference in neutral flux may account for the difference in yield data given in Fig. 3.

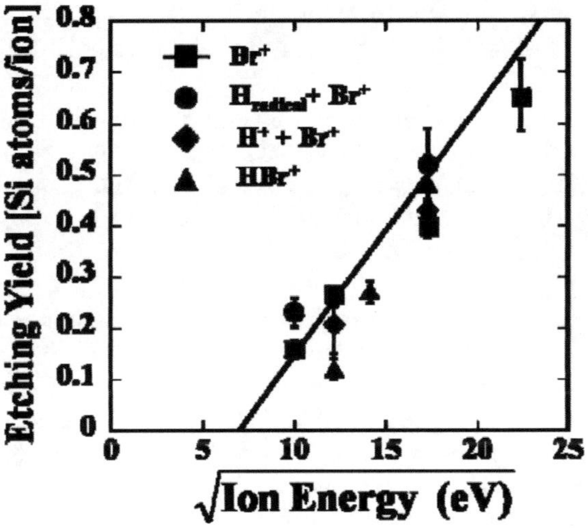

Figure 3. Etching yields of Br ion injections into poly Si under various conditions. The solid line is the fitting curve to the Br^+ injection data denoted by filled squares.

3.2 Etching by HBr^+ or Br^+ ion irradiation

Sputtering yields of various bromine-based etching processes of poly Si are shown in Fig. 3. The yield data for Br^+ ion injections are denoted by filled squares and the linear fit to the data with respect to the square root of the ion injection energy is given by the solid line. The data for HBr^+ ions injections are also shown in Fig. 3, given by the filled triangles. As shown in Fig. 3, the data are somewhat scattered and no clear difference in etching yield between Br^+ and HBr^+ injections has been observed in our study. We also note that the yields shown in Fig. 3 are far smaller than the yield data given in Ref. [1], in which, for example, the poly Si sputtering yield by a Br_2 plasma beam at an injection energy of 100 eV is shown to be approximately 1.

3.3 Simultaneous irradiation of low-energy H radicals and energetic Br^+ ions

The sputtering yield data for simultaneous injections of low-energy charge-neutral H atoms and energetic Br^+ ions (denoted as "$H_{radical} + Br^+$") are given by filled circles in Fig. 3. As mentioned earlier, the atomic hydrogen flux is about 1 ~ 10 times larger than the Br^+ ion flux and therefore Br^+ ions enter the Si surface that is considered to be mostly covered with H radicals in these processes. Yields of spontaneous etching of poly Si by H atoms are known to be negligibly small compared with the sputtering yields obtained here. The data of Fig. 3 suggest that simultaneous injection of low-energy H radicals has little effect on sputtering yields of Br^+ ion injections into poly Si.

3.4 500eV H^+ ion irradiation prior to Br^+ ion irradiation

In actual Si RIE processes by HBr plasmas, energetic H^+ ions that enter the substrate can damage the surface and subsurface region of the substrate. Such damages may affect the rates of reactions between incoming Br^+ ions and substrate poly Si. To check this possibility, we irradiated a poly Si substrate surface initially by 10^{17} /cm^2 of 500eV H^+ ions and then subsequently by a similar dose of Br^+ ions. The sputtering yields obtained from such processes are given by filled diamonds (denoted by "H^++Br^+") in Fig. 3. Given the uncertainty associated with the data shown in Fig. 3, there seems no clear effect of 500eV H^+ ion injections prior to Br^+ ion injections on the Si sputtering yields.

4. Conclusions

In this study we have evaluated sputtering yields of poly Si by Cl^+, Br^+, and HBr^+ ion injections, using a multi-beam system with a mass-selected ion beam injector. Due to the high vacuum (with the typical gas pressure in the reaction chamber being at 10^{-10} Torr during the beam injections) of the beam system, we can eliminate the effects of background gas adsorption on the sample surface. To add effects of neutral gas adsorption on the sample surface, on the other hand, we have used a low-energy neutral beam injection system that can be controlled separately from the ion beam injection system.

Since one of the goals of this study is to clarify effects of hydrogen in HBr RIE processes, we have also evaluated Si yields by Br^+ ions simultaneously injected with H atoms. Also obtained have been sputtering yields by Br^+ ion injections of Si substrates that were previously bombarded by 500eV H^+ ions.

These cases were examined because of a possibility that such atomic or ionic hydrogen can modify the sample surface and therefore affect the sputtering yields by Br^+ ion injections. The yield data obtained in this study, summarized in Fig. 3, have shown, however, that sputtering yields by Br^+ beams, HBr^+ beams, simultaneous injections of H radicals and Br^+ beams, and successive injections of H^+ and Br^+ beams, take more or less the same values as functions of ion beam injection energy. In other words, hydrogen introduced in these processes has little effect on Si sputtering yields by Br^+ ion injections.

It has been long speculated that hydrogen of HBr plasmas enhances Si etching by Br^+ ion bombardment. However our experimental observations above have shown otherwise. The Si yield data given in Ref. [1], which show that the Si yields by HBr plasmas are typically twice larger than those by Br_2 plasmas and the yield values are much larger than those given in Fig. 3, were obtained from plasma beams that contain neutral fluxes much larger than the ionic fluxes. Therefore the difference in yield between HBr plasmas and Br_2 plasmas given in Ref. [1] may reflect the effects of charge-neutral Br atoms that passivate the Si surface, rather than hydrogen radicals or ions. Passivation effects of neutral Br on Br^+ ion etching are beyond the scope of the present work and deferred to future study.

Acknowledgements

This work was partially supported by a Grant-in-Aid for Scientific Research from the Ministry of Education (MEXT) of Japan.

References

[1] S. A. Vitale, H. Chae and H. H. Sawin, J. Vac. Sci. Technol. A**19**, 2197 (2001).
[2] K. Karahashi, K. Yanai, K. Ishikawa, H. Tsuboi, K. Kurihara, M. Nakamura, J.Vac. Sci.Technol. A**22** , 1166 (2004).
[3] J. P. Chang and H. Sawin, J. Vac. Sci. Technol. A**5**, 610 (1997).
[4] S. Tachi and S. Okudaira, J. Vac. Sci. Technol. B**4**, 459 (1986).
[5] C. Steinbrüchel, Appl. Phys. Lett. **55**, 1960 (1989).

Particle-in-Cell Simulations of High-Pressure Hydrogen Plasmas Driven by Nano-second Pulsed High-Voltages

Chieh-Wen Lo[1,2] and Satoshi Hamaguchi[1]

[1] Center of Atomic and Molecular Technologies, Osaka University, Osaka, Japan
[2] Department of Physics, National Cheng Kung University, Tainan, Taiwan

E-mail: cwlo@ppl.eng.osaka-u.ac.jp

Abstract. Particle-in-cell simulations have been performed to clarify the fundamental physical processes of hydrogen plasmas driven by a nano-second pulsed high voltage under a gas pressure near atmospheric pressure. The discharge is a capacitively-coupled plasma generated between metal electrodes whose gap length is about 1 mm. The simulation has shown details of electron and ion dynamics of the hydrogen plasmas during its formation and decay processes. Temporal evolution of the electron energy distribution function has revealed that, due to the fast dynamics, electrons are not in local thermal equilibrium at an early stage of the discharge.

1. Introduction

A hydrogen molecule is one of the simplest diatomic molecules and its atomic processes have been widely studied and generally considered to be well understood. A gas discharge based on pure hydrogen has been also widely studied partly because it is one of the simplest "molecular" gas discharges and also partly because discharges based on its isotopes, i.e., deuterium and tritium, have been considered as plasmas to be used in magnetically-confined thermonuclear fusion reactors such as tokamaks. Hydrogen plasmas are also chemically reactive. For example, hydrogen plasmas are used to clean surfaces of tokamak inner walls or microelectronic devices because highly reactive species in hydrogen plasmas can react with impurities deposited on those surfaces.

Conventionally, for processing applications, hydrogen plasmas are generated under low pressure conditions and therefore materials that can be treated by such hydrogen plasmas have been limited to those that can be placed in a vacuum chamber. On the other hand, for chemically reactive plasmas with little thermal load that can be operated in atmospheric-pressure conditions, the range of their process applications is much wider. For example, a material that contains sufficient amount of water (such as living tissues) can be treated by such plasmas.

In this work, in an attempt to develop plasma processing techniques in atmospheric pressure conditions, we study fundamental properties of atmospheric-pressure plasmas. Especially of interest here is to understand fundamental physical processes that take place during the formation and decay phases of a hydrogen discharge generated by a fast high-voltage pulse. Such pulsed discharges are commonly employed among almost most low-gas-temperature atmospheric-pressure plasmas such as dielectric barrier discharges.

In what follows, we study a (spatially) one-dimensional plasma generated between flat metal electrodes by nano-second pulsed high-voltages at a gas-pressure near atmospheric pressure[1], using numerical simulations. The discharge gas is hydrogen. Physical processes involved in the formation of such plasma are very fast and transient. Therefore, although the pressure is high and the plasma can be very collisional, there is a possibility that the electrons may not be locally in thermal equilibrium. Therefore, as we shall explain more in detail in the following sections, we use a particle method for numerical simulations, from which temporal evolution of the electron energy distribution during the discharge can be obtained.

2. Particle-in-Cell simulation

In this work, a 1d2v (one dimensional in physical space and two dimensional in velocity space) electrostatic particle-in-cell (PIC)[2-4] code with Monte Carlo Collisions (MCC)[2,5] has been developed to simulate capacitively coupled plasma (CCP) discharges. The null collision operator[6] has been employed to improve computational efficiency for collision calculations. This code has been fully tested and verified for sample simulations of several simple cases such as Langmuir oscillation, plasma sheath formation[7], and low pressure radio-frequency (RF) argon discharges[8,9].

3. The numerical model

In the simulation, a hydrogen plasma is generated between two parallel electrodes, i.e., a grounded anode and a cathode connected with an external voltage source. The applied voltage is a triangular pulse with a peak value of -2 kV, and its duration is 10 ns. Its temporal profile is shown in figure 1. The total process simulated in this study lasts for about 25 ns. The distance between the two parallel electrodes is 1.2 mm. The hydrogen gas pressure is set at 0.3 atm. These conditions are similar to those used for the experiments given in Ref. [1].

Initially a low-density plasma is uniformly distributed between the electrodes, which may be considered as a remnant of the previous pulsed discharge. As the initial conditions, the plasma is assumed to be in thermal equilibrium at 300 K. The secondary electron emission coefficient is assumed to be null for the sake of simplicity.

The dominant product of electron impact ionization of a hydrogen molecule is an H_2^+ ion[10]. However, typically the produced H_2^+ ion readily collides with another hydrogen molecule and turns into an H_3^+ ion[11]. Therefore, the only ionic species used in the simulations presented here is H_3^+. The electron-neutral collision processes considered here are elastic collision, ionization to H_2^+ (which is considered to turn to H_3^+ immediately), vibrational excitation of $v = 0 \rightarrow 1$, and electronic excitations to the $b\ {}^3\Sigma^+_u$, $B\ {}^1\Sigma^+_u$, and $E,F\ {}^1\Sigma^+_g$ states. The corresponding cross section data are adopted from Ref. [10] and summarized in figure 2. The ion-neutral elastic collision has been also taken into account and the cross section data for this process have been obtained from Ref. [11].

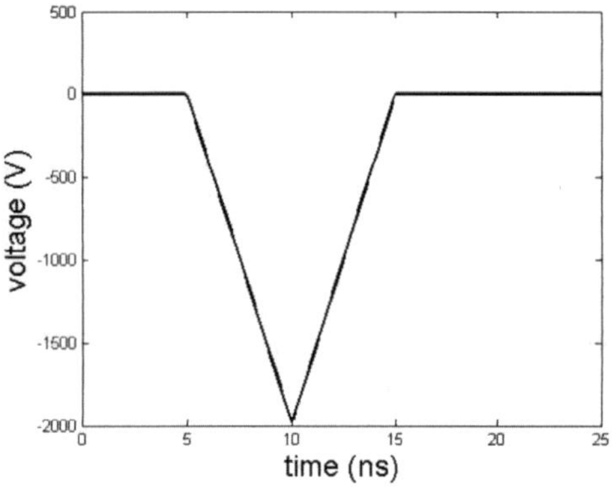

Figure 1. The temporal profile of applied voltage on the cathode.

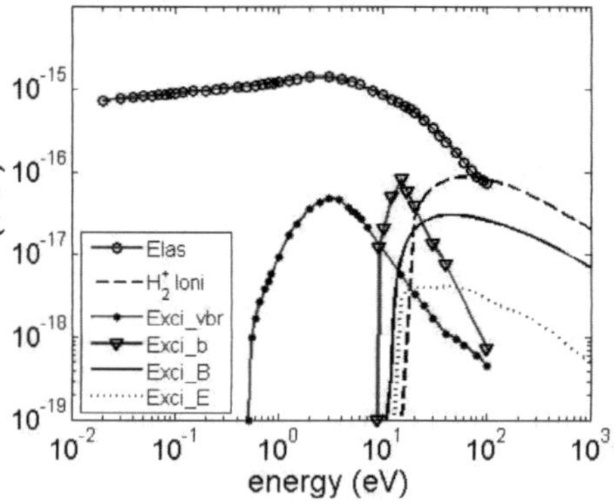

Figure 2. Electron-H_2 molecule collision cross sections adopted from Ref. [10].

4. Simulation results and discussion

Trajectories of charged particles and the time evolution of electric field profile have been evaluated under the condition that the voltage given in figure 1 is applied to the cathode. Details of behavior and statistical characteristics of the hydrogen plasmas obtained from the simulation are described below.

4.1. Electric field and potential profiles

The time evolution of electric field is shown in figure 3(a). At t = 8 ns (i.e., 3 ns after the onset of voltage application) when the magnitude of the applied voltage has risen to 1.2 kV, the electric field profile is fairly flat, which indicates that there is little space charge in the plasma. This is also seen in the electric field potential shown in figure 3(b), where the potential profile is linear. By the time the voltage peaks (at t = 10 ns), however, a thick cathode sheath is formed and the electric field in the bulk is shielded to the high degree. Since no secondary electron emission is taken into this simulation, few electrons remain near the cathode by this time and the plasma is located closer to the anode. After the applied voltage peaks, it is seen that the anode sheath is gradually formed since many electrons are absorbed by the anode until the ambipolar diffusion prevents electrons from further drifting toward the anode.

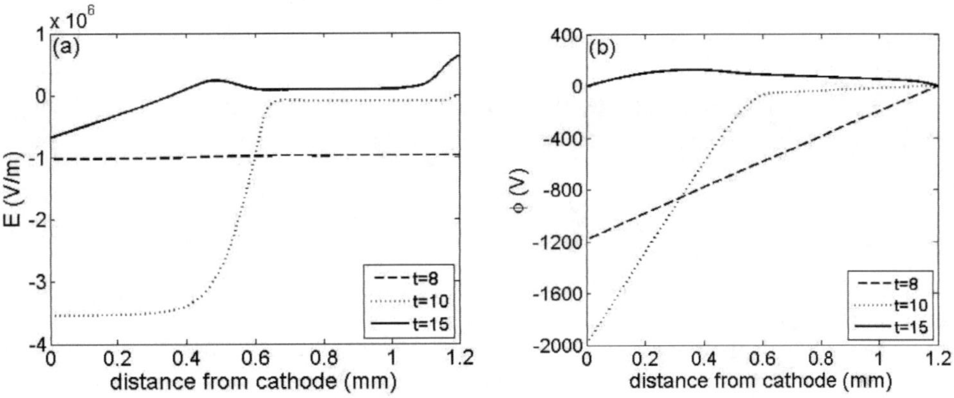

Figure 3. Snapshots of electric field (a) and potential (b) profiles at different time instances. Plasma starts to shield the applied voltage at 8 ns and leads to a plasma potential confining electrons inside the bulk plasma.

4.2. Dynamics of charged particles

Time evolutions of the electron and ion density profiles are given in figure 4. Before the onset of driving voltage application, electrons and ions remain uniform in space with little loss to the electrodes. After the voltage is applied and its magnitude increases, electrons move toward the anode, generating more charged particles through the ionization process along their way toward the anode whereas ions in the system hardly move in this fast time scale due to their large inertia. The number of charged particles thus greatly increases during the period from t = 7 ns to 9 ns as ionizations mostly are occurred in this period. By this time a plasma bulk is distinctly formed with a small peak at the edge close to cathode. The peak is likely to be created by electron impact ionization near the sheath edge where colliding electrons have high kinetic energies due to the acceleration by the strong sheath potential. Since the electrons drift toward the anode with high velocities near the sheath edge whereas ions are nearly immobile in this time scale, the peak in electron density profile is smaller and closer to the anode compared with that in the ion density profile. The cumulated charged particles will gradually diffuse outward after the peak period of ionization rate.

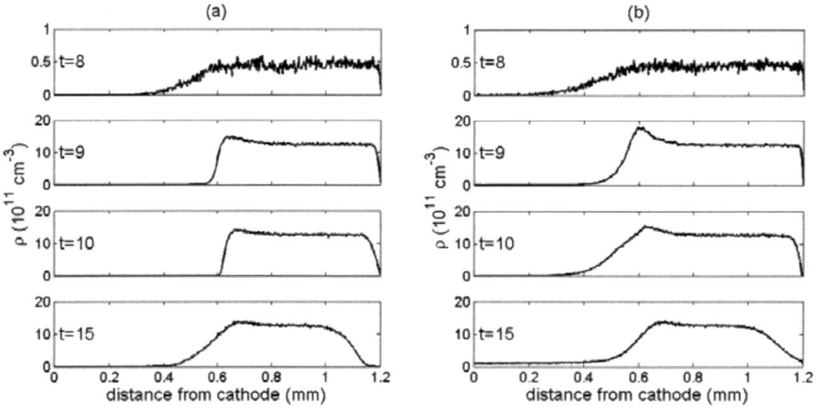

Figure 4. Snapshots of the electron (a) and ion (b) density profiles at different time instances. When the applied voltage is sufficiently high, a sharp peak appears near the cathode sheath edge in the electron or ion distribution as shown at $t = 9$ ns.

4.3. Electron energy probability function

The electron energy distribution function (EEPF)[7] is defined as

$$f_p(E) = \rho_e f_e(E) / \sqrt{E}$$

where E is the electron energy, ρ_e is the electron number density, and $f_e\ (E)$ is the electron energy distribution function whose integration over the entire E yields unity. The EEPF $f_e\ (E)$ is defined at each location. Note that if the electron velocity distribution function is Maxwellian, the logarithm of EEEP, i.e., $\ln f_e\ (E)$ is a linear function of E. In this study, the EEEP was measured near the center of the discharge gap at some time instances, which are shown in Fig. 5. It is seen that, at an early stage of discharge, the EEPF deviates from a simple Maxwellian to a non-Maxwellian due to the electron acceleration by the strong electric field induced by the pulse voltage. Detail analysis of the EEPF time evolution is beyond the scope of present work and deferred to a future study.

5. Conclusions

The dynamics of a hydrogen plasma generated by a high voltage pulse applied to a pair of parallel electrodes has been investigated. It has been found that the formation of the plasma is essentially driven by fast electron dynamics whereas ions and neutral species hardly move in this time scale. Macroscopic dynamics of the discharge obtained from the simulations have been found similar to the recent experimental observations of pulsed hydrogen discharges[1] performed under the conditions nearly identical to those of this study. The simulations have also shown that, despite the high collisionality, electrons in the bulk, measured at the center of the discharge region, can transiently exhibit non-Maxwellian energy distributions. In other words, in such a fast discharge process, the system is not necessarily in local thermal equilibrium.

Acknowledgements

One of the authors (S. H.) appreciates useful discussions with Prof. T. Ito of Osaka University and Prof. U. Czarnetzki of Ruhr University of Bochum on experimental observations given in Ref. [1].

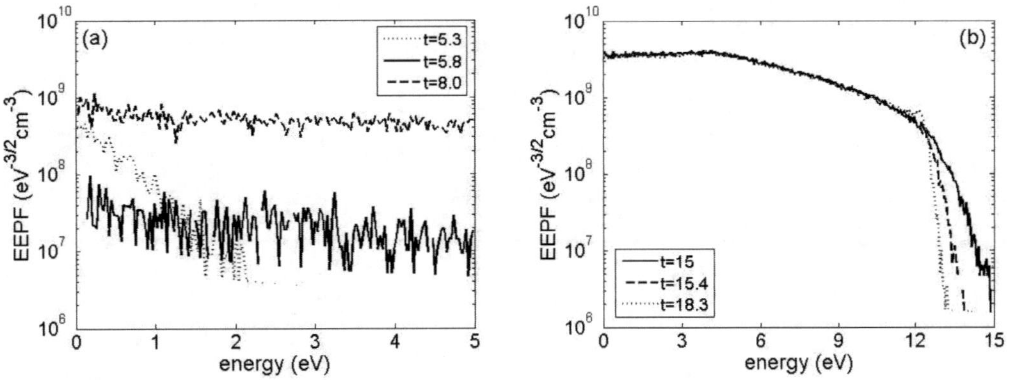

Figure 5. The EEPFs measured in the middle of the discharge at different time instances before (a) and after (b) the peak of the applied voltage.

References

[1] T. Ito, K. Kobayashi, U. Czarnetzki, and S. Hamaguchi, J. Phys. D: Appl. Phys. (2010), *in press*.

[2] C. K. Birdsall and A. B. Langdon, *"Plasma Physics via Computer Simulation,"* McGraw-Hill, New York (1985).

[3] C. K. Birdsall, IEEE Trans. on Plasma Sci. **19** (1991), 65.

[4] J. Dawson, Rev. Modern Phys. **55** (1983), 403.

[5] V. Vahedi and M. Surendra, Comp. Phys. Communication **8**7 (1995), 179.

[6] S. L. Lin and J.N. Bardsley, J Chem. Phys. **66** (1977), 435.

[7] M. A. Lieberman and A. J. Lichtenberg, "Principles of Plasma Discharges and Material Processing", 2nd. Ed., Wiely-Interscience (2005).

[8] V. Vahedi, *et. al*, Plasma Source Sci. Technol. **2** (1993), 261, 273.

[9] D. Vender and R.W. Boswell, IEEE Trans. on Plasma Sci. **18** (1990), 725.

[10] J.-S. Yoon, M.-Y. Song, J.-M Han, S.-H. Hwang, W.-S. Chang, B.-J. Lee, and Y. Itikawa, J. Phys. Chem. Ref. Data **37** (2008), 913.

[11] A.V. Phelps, J. Phys. Chem. Ref. Data **19** (1990), 653.

AUTHOR INDEX